Hydrophobic and Superhydrophobic Organic-Inorganic Nanohybrids

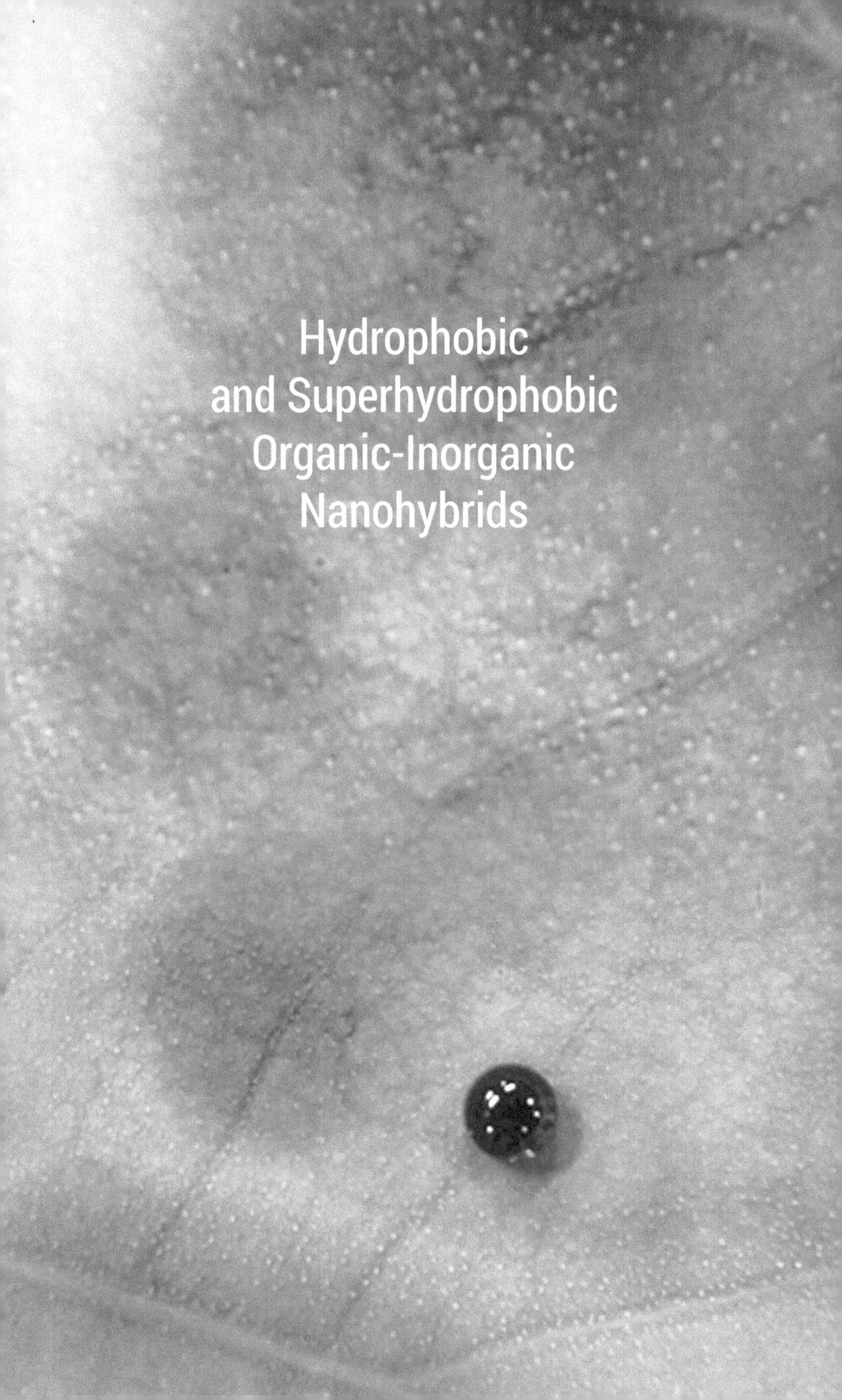

Hydrophobic and Superhydrophobic Organic-Inorganic Nanohybrids

Chang-Sik Ha
Saravanan Nagappan

Pan Stanford Publishing

Published by

Pan Stanford Publishing Pte. Ltd.
Penthouse Level, Suntec Tower 3
8 Temasek Boulevard
Singapore 038988

Email: editorial@panstanford.com
Web: www.panstanford.com

British Library Cataloguing-in-Publication Data
A catalogue record for this book is available from the British Library.

Hydrophobic and Superhydrophobic Organic-Inorganic Nanohybrids

ISBN 978-981-4774-68-0 (Hardcover)
ISBN 978-1-351-20607-5 (eBook)

Contents

Preface

Organic-inorganic hybrid materials are used in various applications because of the presence of dual, enhanced chemical, thermal, and mechanical properties of organic and inorganic materials in a single material. Hybrid materials can be created by combining either an inorganic source to an organic material or an organic source to an inorganic material. In both ways, the material's properties can be improved. Enhanced hybrid materials possess more technical advantages compared to single organic or inorganic materials. Their surface properties can be classified into superhydrophilic, hydrophilic, hydrophobic, or superhydrophobic properties, depending on their surface nature and adhesion performance in relation to water (surface tension of water = 72.0 mN/m).

The technical advantages and some potential applications of organic-inorganic hybrid materials have already been covered by several scientific papers, reviews, and books. Our book, however, exclusively covers hydrophobic and superhydrophobic surfaces based on organic-inorganic nanohybrids, their synthesis and fabrication, and their recent and potential applications in various fields. The book will be a valuable guide for graduate students and scientists who have a background in chemistry, chemical engineering, materials science and engineering, nanotechnology, surface science and engineering, and industrial coating applications.

We would like to express our sincere gratitude to Pan Stanford Publishing for offering us the opportunity to publish this book. We would also like to acknowledge the support given by our former and present researchers in the Nano-Information Materials Laboratory, Pusan National University, Republic of Korea, in bringing out this book.

Chang-Sik Ha and Saravanan Nagappan

Pusan National University

2018

Chapter 1

Hybrid Materials and Surfaces

1.1 Introduction

In materials chemistry, hybrid systems have become popular due to their enhanced properties compared to their individual components. The term "hybrid" refers to the "combination of two or more components into a single domain that reflects the properties of each material in the final material" [1]. Kickelbick et al. provided a more specific definition of a hybrid material. In general, hybrid materials consist of a minimum of two components, such as inorganic and organic materials, which are dispersed molecularly in the material [2].

Hybrid materials are found in one of two forms: homogeneous or heterogeneous. Homogeneous hybrid materials show a combination of monomers and miscible organic and inorganic components, whereas in the heterogeneous form, also known as nanocomposites, one component of the hybrid materials has dimensions ranging from a few angstroms to several nanometers [3]. Hybrid materials are classified further into two classes on the basis of the nature of the interface [3–6]. The interface of a material connected to the organic or inorganic components by weak chemical bonds, such as hydrogen, van der Waals, or ionic bonds, is classified as a class I material, whereas the interface of a material connected to organic or inorganic components by strong chemical bonds, such as covalent

Hydrophobic and Superhydrophobic Organic-Inorganic Nanohybrids
Chang-Sik Ha and Saravanan Nagappan

ISBN 978-981-4774-68-0 (Hardcover), 978-1-351-20607-5 (eBook)
www.panstanford.com

Table 1.1 Comparison of properties of organic-inorganic materials

Properties	Organics (polymers)	Inorganics (SiO_2, TMO)
Nature of bonds	Covalent (C–C) (weaker van der Waals of H bonding)	Ionic or iono-covalent (M-O)
Refractive index	1.2–1.6	1.15–2.7
Density	0.9–1.2	2.0–4.0
Thermal stability	Low (<350°C, except polyimide, 450°C)	High (>>100°C)
T_g (glass transition)	Low (–100°C to 200°C)	High (>200°C)
Electronic properties	Insulating to conductive redox properties	Insulating to semiconductors (SiO_2, TMO), redox properties, magnetic properties
Mechanical properties	Elasticity, plasticity, rubbery (depending on T_g)	Hardness, strength, fragility
Processibility	High: molding, casting, machining thin films from solution control of the viscosity	Low for powders (need to be mixed with polymers or dispersed in solutions); high for sol-gel coatings (similar to polymers)
Hydrophobicity, permeability	Hydrophilic, hydrophobic ± permeable to gases	Hydrophilic, low permeability to gases

TMO, transition metal oxide.
Source: Adapted from Ref. [4] with permission from The Royal Society of Chemistry.

Wen et al. explained the different approaches of the synthesis of organic-inorganic hybrid materials by the sol-gel method using various starting components [22]. The synthetic approaches of organic-inorganic hybrid materials are as follows. One or more low-molecular-weight organoalkoxysilane precursors are used for the synthesis of hybrid materials, whereas the organic groups are present within the inorganic network. Similarly, polymers or oligomers are also used as an organic source with the inorganic component to synthesize organic-inorganic hybrid materials. Hybrid materials are also synthesized by interpenetrating polymer networks (IPNs) and the simultaneous formation of inorganic and organic phases, impregnation or entrapping organic component (guest) within the inorganic gel matrixes (host), and in-situ formation of inorganic species within a polymer matrix. The optical, electrical, and flexible properties were improved by the introduction of an organic phase in the inorganic component. Moreover, the synthesized hybrid materials can be used for a range of applications, such as self-healing, antifouling, and anticorrosive coatings; adhesive and contact lenses; reinforcement of plastic and elastomers; abrasion- and scratch-resistant coatings' sensors; catalysis; and other applications [22].

In recent decades, mesoporous nanoparticles (MNPs) and periodic mesoporous organosilica (PMO) synthesized from various organic and inorganic components have shown excellent properties for a range of applications, particularly biomedical, energy, and environmental applications [23]. Mesoporous materials are synthesized generally by the cohydrolysis and condensation of tetraalkoxysilanes [$Si(OR)_4$] in the presence of water, a catalyst, and a surfactant (as a structure-directing agent [SDA]).

Various types of surfactants are used for the synthesis of mesoporous materials. Moreover, the porous structure of the material depends on the type of surfactant used for the synthesis. The mesoporous structure was developed from the synthesized materials by either thermal or chemical treatments. The mesoporous organic-inorganic hybrid materials were obtained by surface functionalization of the synthesized mesoporous material with a range of organic precursors.

Athens et al. reviewed functionalized mesoporous organic-inorganic hybrid materials with various precursors [24]. In most cases, the terminal organosilanes react with the mesoporous materials to produce the majority of the organic moieties at the mesoporous surface [24]. The functionalized mesoporous materials are used for many applications based on surface functional groups, such as metal and dye adsorption, drug delivery, and catalysis [25]. PMO is a type of organic-inorganic hybrid material that is synthesized generally by the hydrolysis and condensation of mono- or bis-silylated precursors (Fig. 1.1). The inorganic pore walls in PMOs are occupied more uniformly by the organic moieties. The obtained PMOs can show a more hydrophobic and less brittle nature [24, 26, 27]. Hoffmann et al. briefly reviewed the synthesis pathways of silica-based mesoporous organic-inorganic hybrid materials. In particular, they focused on the synthesis of PMOs with various SDAs or surfactants and precursors used for the synthesis of PMOs (Table 1.2) [27–41].

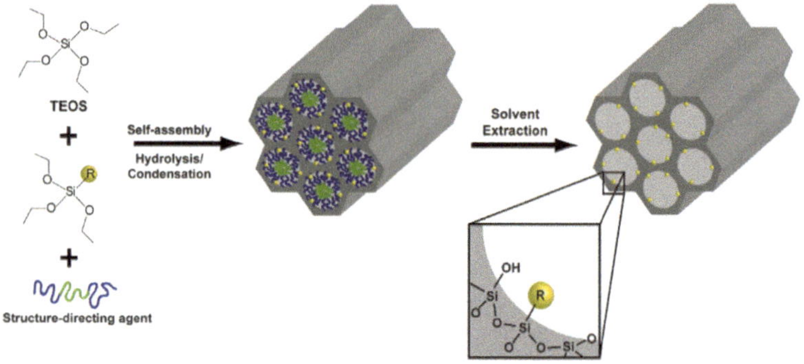

Figure 1.1 Co-condensation approach to the functionalization of mesoporous inorganic materials by direct coassembly and incorporation of the functional moiety (R) species during synthesis. Reprinted from Ref. [24]. Copyright (2009), with permission from Elsevier.

The continuous growth in the development of new types of organic-inorganic hybrid materials leads to a further drive to diversify the field to other applications. The synthesis of functional hybrid materials is another important characteristic to improve the chemical and physical properties of hybrid materials, such as mechanical, thermal, electrical, magnetic, optical, and biological activities. The development of a multifunctional nanohybrid materials by the combination of atomic or nanosized inorganic and organic or bioactive molecules into a single material has also focused considerable attention in many applications. More recently, organic-inorganic hybrid materials prepared by the formation of a zeolite-like metal organic framework (MOF) structure showed diverse use in catalysis, energy, and environmental applications [42–45]. Millini et al. briefly reviewed three main approaches for the synthesis of hybrid organic-inorganic zeolites: (i) incorporation of an organic moiety into a known zeolite framework, (ii) the pillaring of preformed layered zeolite precursors, and (iii) direct synthesis of microporous crystalline hybrid organic-inorganic metallosilicates [42].

Table 1.2 Various precursors and surfactants used for the synthesis of mesoporous nanoparticles and periodic mesoporous organosilicas [28–41]

Precursors for MNPs	Surfactant	Ref.	Precursors for PMOs	Surfactant	Ref.
$Si(OR)_4$	CTAC	27	1,4-bis(triethoxysilyl) benzene	Pluronic P123	32
$SiCl_4$	CTAB	28	Bridged silsesquioxane	Brij 56	33
$Ti(OR)_4$	Pluronic P123	29	4,4′bis(triethoxysilyl) biphenyl	OTAC	34
$TiCl_4$	Synperonic F108	29	Phenyl sulfide bridge-bonded silsesquioxane	Brij 76	35
$Zr(OR)_4$	Synperonic F108	30	Bis-(triethoxysila) methane	OTAC	36
$ZrCl_4$	Pluronic P123	31	Bis(3-trimethoxysilyl) propylamine (BTMSPA)	CTAB/Brij 30	37
$Al(OR)_3$	Synperonic F108	29	Bis(triethoxysilyl) ethane (BTESE)	CTAC	38
$AlCl_3$	Pluronic P123	29	Cyclam moiety–based silane precursor	Pluronic P123	39
WCl_6	Pluronic P123	29	4,4A-bipyridine-based silane precursor	CTAB	40
Nb_2O_5	Pluronic P123	30	–	–	–

CTAC, cetyltrimethyl ammonium chloride; CTAB, cetyltrimethyl ammonium bromide; OTAC, octadecyltrimethyl ammonium chloride; Pluronic P123 and synperonic (F108), poly(ethylene glycol)-poly(propylene glycol)-poly(ethylene glycol) triblock copolymers; Brij 30, polyoxyethylene (4) lauryl ether; Brij 56, polyoxyethylene (10) cetyl ether; Brij 76, polyoxyethylene (10) stearyl ether.

Potier et al. synthesized a novel organic-inorganic copolymer with self-healing properties using n-butyl acrylate and tin oxo-cluster in the presence of azoisobutyronitrile (AIBN) [46]. The self-healing nature of the copolymer was attributed to the development of a non-covalent interaction (ionic bonds) in the hybrid interfaces, which lead to the formation of crosslinking with the polymer. Moreover, the polymer exhibited elastic behavior and the original structure could be reformed even under severe mechanical damage [46]. The easy curable organic-inorganic hybrid materials synthesized using UV-curable resin showed excellent performance in several coating applications [47]. The increase in the number of publications and citations (from articles and reviews) over the last 10 years under the topic of organic-inorganic hybrid materials shows the current trends of the synthesis of organic-inorganic hybrid materials that continuously emerge for many applications (Fig. 1.2).

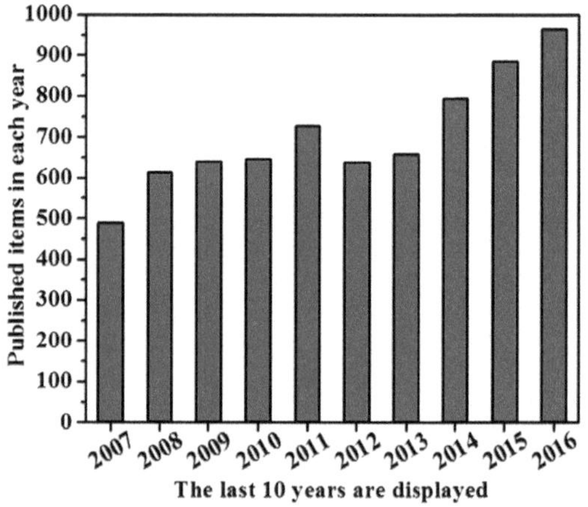

Figure 1.2 Number of items published in each year under the topic "organic-inorganic hybrid materials." *Source*: Web of Science^{TM}; keyword: organic-inorganic hybrid materials (only articles and reviews).

1.3 Surface Wettability

The surface wettability on a solid surface has attracted considerable attention in various fields due to the adhesion behavior of a water droplet on the substrate surface. The surface wettability is determined by measuring the surface contact angle (CA) using a CA analyzer. In general, water (surface tension 72.8 mN/m) is used to measure the surface CA. The surface wettability of a material is classified on the basis of the wettability of the solid surface to a water droplet. The wettability is also dependent on the surface tension of a material. A surface with high or low surface tension can be hydrophilic or hydrophobic according to the surface tension of the liquid. Table 1.3 lists the surface tension of various common liquids and polymers used in a chemical laboratory. The surface tension of a solid material can be calculated by the measuring the CA between a solid surface and a droplet of liquid on the surface. The surface wettability on the solid surface depends on several factors, such as micronanoroughness, homogeneous or heterogeneous mixing of materials, chemical composition, surface tension and surface energy, and adhesion of the coated materials to the solid surface [48–53].

Table 1.3 Surface tension of common liquids and polymers used in a chemical laboratory

Liquids and polymers	Surface tension (mN/m)
Trifluoroacetic acid	13.63 (24°C)
Hexane	17.91 (25°C)
Iso-octane	18.77
Heptane	20.30
Triethylamine	20.66
Isopropyl alcohol	21.79 (15°C)
Ethyl alcohol	22.32
Methanol	22.55
Acetone	23.32
Cyclohexane	24.98
Tetrahydrofuran	26.40 (25°C)
Chloroform	27.16
Toluene	28.53
Chlorobenzene	33.28
N,N-dimethylformamide	36.76
Pyridine	36.88
Water	72.80
Polytetrafluoroethylene	18
Polyvinylidiene fluoride	25
Polypropylene	29
Polyethylene	31
Polystyrene	33
Polyvinyl alcohol	37
Polyvinyl chloride	39
Starch	39
Cellulose	44

150° [52, 53, 64, 65]. Superhydrophobic surfaces are inspired from leaves and flowers (lotus, tree of heaven, rice leaves, and rose petals), and animal species (water strider leg, shark skin, spider silks, cicada wings, mosquito eyes, butterfly wings, gecko feet, duck feather, peacock feather, and beetle shells) [65–69].

Superhydrophobic materials and surfaces are important from both theoretical and practical points of view because of their easy fabrication on a wide range of substrates, which can be useful in several applications, such as selective oil and water absorption and separation, selective metal ion adsorption, controlled drug delivery, antibacterial and anticorrosion, cell adhesion and culture, solar cells, displays, flexible and colorful substrate fabrication, micronanopatterning, microfluidics, sensors, self-healing, self-cleaning, and antireflective coatings [65–82]. Switchable superhydrophobic surfaces can also be obtained using various stimuli-responsive materials. The surface properties can change from superhydrophobic to hydrophobic or hydrophilicity under pH, temperature, UV, plasma, electron irradiation, lithography, and laser, which can be switched back to superhydrophobic properties in the absence of the above source [83]. Switchable superhydrophobic surfaces are used widely for sensors as well as for biological, antifogging, and antifouling applications [84].

Recently, organic-inorganic hybrid materials have also attracted attention in the fabrication of hydrophobic and superhydrophobic surfaces due to the resemblance of excellent properties that can be used broadly for a range of applications in several fields [85]. In the recent few decades, hybrid materials made from organic-inorganic materials have been used widely in many applications particularly in electronics, medical, specialty coatings, automobiles, and textiles.

The research articles and review papers published over the past 10 years under the topic "hydrophobic and superhydrophobic surfaces" also highlighted the continuous development of the fields for diverse applications (Figs. 1.3 and 1.4). This book explains briefly the various routes of the synthesis and fabrication of hydrophobic and superhydrophobic organic-inorganic hybrid materials and surfaces, as well as their many applications.

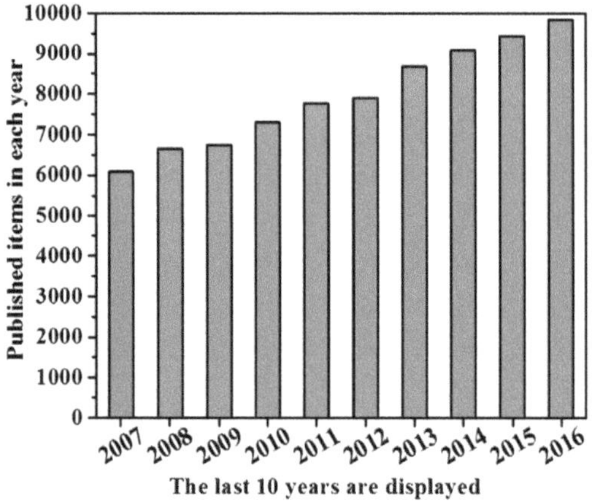

Figure 1.3 Number of items published in each year under the topic "hydrophobic." *Source*: Web of Science^{TM}; keyword: hydrophobic (only articles and reviews).

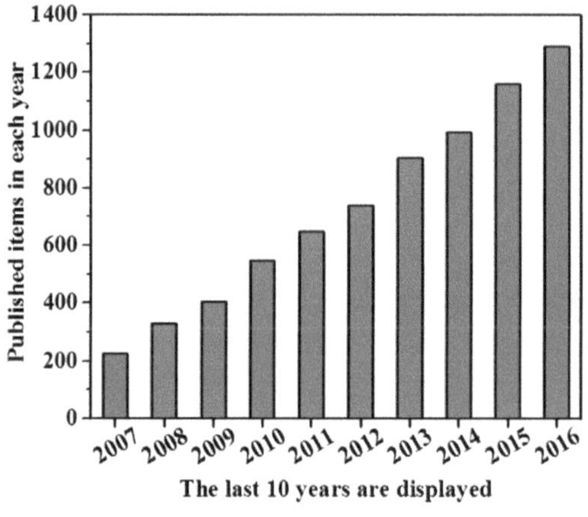

Figure 1.4 Number of items published in each year under the topic "superhydrophobic." *Source*: Web of Science^{TM}; keyword: superhydrophobic (only articles and reviews).

References

1. Ashby, M. F., and Bréchet, Y. J. M. (2003). Designing hybrid materials, *Acta Mater.*, **51**, pp. 5801–5821.
2. Kickelbick, G. (2014). Hybrid materials: past, present and future, *Hybrid Mater.*, **1**, pp. 39–51.

3. Sanchez, C., and Ribot, F. (1994). Design of hybrid organic-inorganic materials synthesized via sol-gel chemistry, *New J. Chem.*, **18**, pp. 1007–1047.
4. Judeinstein, P., and Sanchez, C. (1996). Hybrid organic-inorganic materials: a land of multidisciplinarity, *J. Mater. Chem.*, **6**, pp. 511–525.
5. Gomez-Romero, P. (2001). Hybrid organic-inorganic materials: in search of synergic activity, *Adv. Mater.*, **13**, pp. 163–174.
6. Sanchez, C., Julián, B., Belleville, P., and Popall, M. (2005). Applications of hybrid organic–inorganic nanocomposites, *J. Mater. Chem.*, **15**, pp. 3559–3592.
7. Unterlass, M. M. (2016). Green synthesis of inorganic-organic hybrid materials: state of the art and future perspectives, *Eur. J. Inorg. Chem.*, **8**, pp. 1135–1156.
8. Chen, Y., Meng, Q., Wu, M., Wang, S., Xu, P., Chen, H., Li, Y., Zhang, L., Wang, L., and Shi, J. (2014). Hollow mesoporous organosilica nanoparticles: a generic intelligent framework-hybridization approach for biomedicine, *J. Am. Chem. Soc.*, **136**, pp. 16326–16334.
9. Stöferle, T., Scherf, U., and Mahrt, R. F. (2009). Energy transfer in hybrid organic/ inorganic nanocomposites, *Nano Lett.*, **9**, pp. 453–456.
10. Briseno, A. L., and Yang, P. (2009). Combining chemical worlds, *Nat. Mater.*, **8**, pp. 7–8.
11. Gomez-Romero, P., Chojak, M., Gallegos, K. C., Asensio, J. A., Kulesza, P. J., Pastor, N. C., and Cantú, M. L. (2003). Hybrid organic–inorganic nanocomposite materials for application in solid state electrochemical supercapacitors, *Electrochem. Commun.*, **5**, pp. 149–153.
12. Puglisi, A., Benaglia, M., Annunziata, R., Chiroli, V., Porta, R., and Gervasini, A. (2013). Chiral hybrid inorganic-organic materials: synthesis, characterization, and application in stereoselective organocatalytic cycloadditions, *J. Org. Chem.*, **78**, pp. 11326–11334.
13. Nguyen, T. P., Lee, C. W., Hassen, S., and Le, H. C. (2009). Hybrid nanocomposites for optical applications, *Solid State Sci.*, **11**, pp. 1810–1814.
14. Kaushik, A., Kumar, R., Arya, S. K., Nair, M., Malhotra, B. D., and Bhansali, S. (2015). Organic-inorganic hybrid nanocomposite-based gas sensors for environmental monitoring, *Chem. Rev.*, **115**, pp. 4571–4606.
15. Kim, Y., Lee, W. K., Cho, W. J., Ha, C. S., Ree, M., and Chang, T. (1997). Morphology of organic–inorganic hybrid composites in thin films as multichip packaging material, *Polym. Int.*, **43**, pp. 129–136.

16. Ogoshi, T., and Chujo, Y. (2005). Organic inorganic polymer hybrids prepared by the sol-gel method, *Compos. Interfaces*, **11**, pp. 539–566.
17. Gomez-Romero, P., and Sanchez, C. (2005). Hybrid materials functional properties from maya blue to 21st century materials, *New J. Chem.*, **29**, pp. 57–58.
18. Mammeri, F., Bourhis, E. L., Rozes, L., and Sanchez, C. (2005). Mechanical properties of hybrid organic–inorganic materials, *J. Mater. Chem.*, **15**, pp. 3787–3811.
19. Soler-Illia, G. J. A. A., Sanchez, C., Lebeau B., and Patarin, J. (2002). Chemical strategies to design textured materials: from microporous and mesoporous oxides to nanonetworks and hierarchical structures, *Chem. Rev.*, **102**, pp. 4093–4138.
20. Sanchez, C., Ribot, F., and Lebeau, B. (1999). Molecular design of hybrid organic-inorganic nanocomposites synthesized via sol-gel chemistry, *J. Mater. Chem.*, **9**, pp. 35–44.
21. Schubert U. (2015). Chemistry and fundamentals of the sol–gel process, in Levy, D. and Zayat M., ed., *The Sol–Gel Handbook: Synthesis, Characterization, and Applications*, 1st edition, Wiley-VCH Verlag GmbH & Co. KGaA., p. 5.
22. Wen, B., and Wilkes, G. L. (1996). Organic/inorganic hybrid network materials by the sol-gel approach, *Chem. Mater.*, **8**, pp. 1667–1681.
23. Liu, R., Shi, Y., Wan, Y., Meng, Y., Zhang, F, Gu, D., Chen, Z., Tu, B., and Zhao, D (2006). Triconstituent co-assembly to ordered mesostructured polymer-silica and carbon-silica nanocomposites and large-pore mesoporous carbons with high surface areas. *J. Am. Chem. Soc.*, **128**, pp. 11652–11662.
24. Athens, G. L., Shayib, R. M., and Chmelka, B. F. (2009). Functionalization of mesostructured inorganic–organic and porous inorganic materials, *Curr. Opin. Colloid Interface Sci.*, **14**, pp. 281–292.
25. Ivanova, V. Y., Salvado, I., Fernandes, M., Vassileva, P., and Georgiev, R. (2013). Silica based organic-inorganic hybrid materials as potential adsorbents for Hg(II) ions, *J. Chem. Technol. Metall.*, **48**, pp. 577–584.
26. Park, S. S., and Ha, C. S. (2006). Organic–inorganic hybrid mesoporous silicas: functionalization, pore size, and morphology control, *Chem. Rec.*, **6**, pp. 32–42.
27. Hoffmann, F., Cornelius, M., Morell, J., and Fröba, M. (2006). Silica-based mesoporous organic–inorganic hybrid materials, *Angew. Chem. Int. Ed.*, **45**, pp. 3216–3251.

28. Qiao, Z. A., Zhang, L., Guo, M., Liu, Y., and Huo, Q. (2009). Synthesis of mesoporous silica nanoparticles via controlled hydrolysis and condensation of silicon alkoxide, *Chem. Mater.*, **21**, pp. 3823–3829.
29. Khushalani, D., Hasenzahl, S., and Manna, S. (2001). Synthesis of mesoporous silica with embedded nickel nanoparticles, *J. Nanosci. Nanotechnol.*, **1**, pp. 129–132.
30. Tian, B., Liu, X., Zhang, Z., Tu, B., and Zhao, D. (2002). Syntheses of high-quality mesoporous materials directed by blends of nonionic amphiphiles under nonaqueous conditions, *J. Solid State Chem.*, **167**, pp. 324–329.
31. Cheng, W., Baudrin, E., Dunn, B., and Zink, J. I. (2001). Synthesis and electrochromic properties of mesoporous tungsten oxide, *J. Mater. Chem.*, **11**, pp. 92–97.
32. Yang, P., Zhao, D., Margolese, D. I., Chmelka, B. F., and Stucky, G. D. (1998). Generalized syntheses of large-pore mesoporous metal oxides with semicrystalline frameworks, *Nature*, **396**, pp. 152–155.
33. Goto, Y., and Inagaki, S. (2002). Synthesis of large-pore phenylene-bridged mesoporous organosilica using triblock copolymer surfactant, *Chem. Commun.*, pp. 2410–2411.
34. Hunks, W. J., and Ozin, G. A. (2004). Periodic mesoporous organosilicas with phenylene bridging groups, 1,4-$(CH_2)_nC_6H_4$(n=0-2), *Chem. Mater.*, **16**, pp. 5465–5472.
35. Kapoor, M. P., Yang, Q., and Inagaki, S. (2002). Self-assembly of biphenylene-bridged hybrid mesoporous solid with molecular-scale periodicity in the pore walls, *J. Am. Chem. Soc.*, **124**, pp. 15176–15177.
36. Hunks, W. J., and Ozin, G. A. (2004). Periodic mesoporous phenylenesilicas with etheror sulfide hinge groups—a new class of PMOs with ligand channels, *Chem. Commun.*, pp. 2426–2427.
37. Landskron, K., and Ozin, G. A. (2004). Periodic mesoporous dendrisilicas, *Science*, **306**, pp. 1529–1532.
38. Wahab, M. A., Imae, I., Kawakami, Y., and Ha, C. S. (2005). Periodic mesoporous organosilica materials incorporating various organic functional groups: synthesis, structural characterization, and morphology, *Chem. Mater.*, **17**, pp. 2165–2174.
39. Burleigh, M. C., Markowitz, M. A., Spector, M. S., and Garber, B. P. (2001). Direct synthesis of periodic mesoporous organosilicas: functional incorporation by co-condensation with organosilanes, *J. Phys. Chem. B*, **105**, pp. 9935–9942.

40. Corriu, R. J. P., Mehdi, A., Reyé, C., and Thieuleux, C. (2002). Mesoporous hybrid materials containing functional organic groups inside both the framework and the channel pores, *Chem. Commun.*, pp. 1382–1383.
41. Alvaro, M., Ferrer, B., Fornes, V., and Garcia, H. (2001). A periodic mesoporous organosilica containing electron acceptor viologen units, *Chem. Commun.*, pp. 2546–2547.
42. Millini, R., and Bellussi, G. (2016). Hybrid organic–inorganic zeolites: status and perspectives, *Catal. Sci. Technol.*, **6**, pp. 2502–2527.
43. Dhakshinamoorthy, A., and Garcia, H. (2012). Catalysis by metal nanoparticles embedded on metal–organic frameworks, *Chem. Soc. Rev.*, **41**, pp. 5262–5284.
44. Wang, C., Liu, X., Demir, N. K., Chen, J. P., and Li, K. (2016). Applications of water stable metal-organic frameworks, *Chem. Soc. Rev.*, **45**, pp. 5107–5134.
45. Li, S., and Huo, F. (2015). Metal–organic framework composites: from fundamentals to applications, *Nanoscale*, **7**, pp. 7482–7501.
46. Potier, F., Guinault, A., Delalande, S., Sanchez, C., Ribotabc, A., and Rozes, L. (2014). Nano-building block based-hybrid organic-inorganic copolymers with self-healing properties, *Polym. Chem.*, **5**, pp. 4474–4479.
47. Malucelli, G. (2016). Hybrid organic/inorganic coatings through dual-cure processes: state of the art and perspectives, *Coatings*, **6**, p. 10.
48. Young, T. (1805). An essay on the cohesion of fluids, *Phil. Trans. R. Soc. Lond.*, **95**, pp. 65–87.
49. Wenzel, R. N. (1936). Resistance of solid surfaces to wetting by water, *Ind. Eng. Chem.*, **28**, pp. 988–994.
50. Extrand, C. W. (1998). A thermodynamic model for contact angle hysteresis, *J. Colloid Interface Sci.*, **207**, pp. 11–19.
51. Extrand, C. W. (2002). Model for contact angles and hysteresis on rough and ultraphobic surfaces, *Langmuir*, **18**, pp. 7991–7999.
52. Cassie, A. B. D., and Baxter, S. (1944). Wettability of porous surfaces, *Trans. Faraday. Soc.*, **40**, pp. 546–551.
53. Cassie, A. B. D. (1948). Contact angles, *Discuss. Faraday Soc.*, **3**, pp. 11–16.
54. Chau, T. T., Bruckard, W. J., Koh, P. T. L., and Nguyen, A. V. (2009). A review of factors that affect contact angle and implications for flotation practice, *Adv. Colloid Interface Sci.*, **150**, pp. 106–115.
55. Quéré, D. (2002). Rough ideas on wetting, *Physica A*, **313**, pp. 32–46.

56. Marmur, A. (2004). The lotus effect: superhydrophobicity and metastability, *Langmuir*, **20**, pp. 3517–3519.

57. de Gennes, P. G. (1985). Wetting: statics and dynamics, *Rev. Mod. Phys.*, **57**, pp. 827–863.

58. Drelich, J., Miller, J. D., and Good, R. J. (1996). The effect of drop (bubble) size on advancing and receding contact angles for heterogeneous and rough solid surfaces as observed with sessile-drop and captive-bubble techniques, *J. Colloid Interface Sci.*, **179**, pp. 37–50.

59. McHale, G., Shirtcliffe, N. J., and Newton, M. I. (2004). Contact-angle hysteresis on super-hydrophobic surfaces, *Langmuir*, **20**, pp. 10146–10149.

60. Schmidt, D. L., Brady, R. F., Lam, K., Schmidt, D. C., and Chaudhury, M. K. (2004). Contact angle hysteresis, adhesion, and marine biofouling, *Langmuir*, **20**, pp. 2830–2836.

61. Yeh, K. Y., Chen, L. J., and Chang, J. Y. (2008). Contact angle hysteresis on regular pillar-like hydrophobic surfaces, *Langmuir*, **24**, pp. 245–251.

62. Nagappan, S., Park, S. S., and Ha, C. S. (2015). Recent advances in superhydrophobic nanomaterials and nanoscale systems, *J. Nanosci. Nanotechnol.*, **14**, pp. 1441–1462.

63. Nagappan, S., and Ha, C. S. (2015). Emerging trends in superhydrophobic surface based magnetic materials: fabrications and their potential applications, *J. Mater. Chem. A*, **3**, pp. 3224–3251.

64. Barthlott W., and Neinhuis, C. (1997). Purity of the sacred lotus, or escape from contamination in biological surfaces, *Planta*, **202**, pp. 1–8.

65. Nagappan, S., Park, J. J., Park, S. S., Lee W. K., and Ha, C. S. (2013). Bio-inspired, multi-purpose and instant superhydrophobic–superoleophilic lotus leaf powder hybrid micro–nanocomposites for selective oil spill captur, *J. Mater. Chem. A*, **1**, pp. 6761–6769.

66. Youngblood, J. P., and Sottos, N. R. (2008). Bio-inspired materials for self-cleaning and self-healing, *MRS Bull.*, **33**, pp. 732–741.

67. Xia, F., and Jiang, L. (2008). Bio-inspired, smart, multiscale interfacial materials, *Adv. Mater.*, **20**, pp. 2842–285.

68. Liu, K., and Jiang, L. (2012). Bio-inspired self-cleaning surfaces, *Annu. Rev. Mater. Res.*, **42**, pp. 231–263.

69. Darmanin, T., and Guittard, F. (2014). Recent advances in the potential applications of bioinspired superhydrophobic materials, *J. Mater. Chem. A*, **2**, pp. 16319–16359.

70. Cheng, M., Gao, Y., Guo, X., Shi, Z., Chen, J. F., and Shi, F. (2011). A functionally integrated device for effective and facile oil spill cleanup, *Langmuir*, **27**, pp. 7371–7375.

71. Zhu, Q., Pan, Q. M., and Liu, F. T. (2011). Facile removal and collection of oils from water surfaces through superhydrophobic and superoleophilic sponges, *J. Phys. Chem. C*, **115**, pp. 17464–17470.

72. Alves, N. M., Shi, J., Oramas, E., Santos, J. L., Tomás, H., and Mano, J. F. (2009). Bioinspired superhydrophobic poly(L-lactic acid) surfaces control bone marrow derived cells adhesion and proliferation, *J. Biomed. Mater. Res. Part A*, **91A**, pp. 480–488.

73. Yohe, S. T., Herrera, V. L., Colson, Y. L., and Grinstaff, M. W. (2012). 3D superhydrophobic electrospun meshes as reinforcement materials for sustained local drug delivery against colorectal cancer cells, *J. Control. Release*, **162**, pp. 92–101.

74. Privett, B. J., Youn, J., Hong, S. A., Lee, J., Han, J. H., Shin, J. H., and Schoenfisch, M. H. (2011). Antibacterial fluorinated silica colloid superhydrophobic surfaces, *Langmuir*, **27**, pp. 9597–9601.

75. Asmatulu, R., Ceylan, M., and Nuraje, N. (2011). Study of superhydrophobic electrospun nanocomposite fibers for energy systems, *Langmuir*, **27**, pp. 504–507.

76. Gokhale, R., Agarkar, S., Debgupta, J., Shinde, D., Lefez, B., Banerjee, A., Jog, J., More, M., Hannoyer, B., and Ogale, S. (2012). Laser synthesized super-hydrophobic conducting carbon with broccoli-type morphology as a counter-electrode for dye sensitized solar cells, *Nanoscale*, **4**, pp. 6730–6734.

77. Tadanaga, K., Morinaga, J., Matsuda, A., and Minami, T. (2000). Superhydrophobic-superhydrophilic micropatterning on flowerlike alumina coating film by the sol–gel method, *Chem. Mater.*, **12**, pp. 590–592.

78. Li, H. L., Wang, J. X., Yang, L. M., and Song, Y. L. (2008). Superoleophilic and superhydrophobic inverse opals for oil sensors, *Adv. Funct. Mater.*, **18**, pp. 3258–3264.

79. Wang, C. F., Chen, W. Y., Cheng, H. Z., and Fu, S. L. (2010). Pressure-proof superhydrophobic films from flexible carbon nanotube/polymer coatings, *J. Phys. Chem. C*, **114**, pp. 15607–15611.

80. Bai, C. (2005). Ascent of nanoscience in China, *Science*, **309**, pp. 61–63.

81. Zhang, F., Zhao, L., Chen, H., Xu, S., Evans, D. G., and Duan, X. (2008). Corrosion resistance of superhydrophobic layered double hydroxide films on aluminum, *Angew. Chem. Int. Ed.*, **47**, pp. 2466–2469.

82. Zhang, Q., He, M., Chen, J., Wang, J., Song, Y., and Jiang, L. (2013). Anti-icing surfaces based on enhanced self-propelled jumping of condensed water microdroplets, *Chem. Commun.*, **49**, pp. 4516–4518.

83. Tian, D. L., Zhang, X. F., Tian, Y., Wu, Y., Wang, X., Zhai, J., and Jiang, L. (2012). Photo-induced water–oil separation based on switchable superhydrophobicity–superhydrophilicity and underwater superoleophobicity of the aligned ZnO nanorod array-coated mesh films, *J. Mater. Chem.*, **22**, pp. 19652–19657.

84. Wang, H. Y., Kwok, D. T. K., Xu, M., Shi, H. G., Wu, Z. W., Zhang, W., and Chu, P. K. (2012). Tailoring of mesenchymal stem cells behavior on plasma-modified polytetrafluoroethylene, *Adv. Mater.*, **24**, pp. 3315–3324.

85. Nagappan, S., and Ha, C. S. (2014). Hydrophobic and superhydrophobic organic-inorganic hybrids and their applications, *Austin J. Chem. Eng.*, **1**, p. 1003. ISSN: 2381–8905.

Chapter 2

Hydrophobic Organic-Inorganic Nanohybrids

2.1 Introduction

Hydrophobicity is important in various coating applications to maintain or enhance the life cycle of the coating product. Hydrophobic materials can show strong hydrophobic interactions at the surface of a material. Hydrophobic surfaces have few interactions with water droplets due to the presence of nonpolar functional groups at the surface. This may show weak intermolecular forces, such as hydrogen bonding or van der Waals interactions, while the water droplet is in contact with the surface. The hydrophobic organic-inorganic nanohybrids have been synthesized by changing the surface functional groups using several hydrophobic precursors. The hydrophobicity of organic-inorganic hybrid materials depends on several factors, such as the types of surface functional groups, surface energy and surface tension, microscopic geometry of the surface, and interaction between the surface of the material and water molecules [1–3].

The strength of the hydrophobic interactions with water on a solid surface is determined by several factors:

- Hydrophobic interactions depend on the number of hydrocarbons attached to the molecule. A molecule with a

Hydrophobic and Superhydrophobic Organic-Inorganic Nanohybrids
Chang-Sik Ha and Saravanan Nagappan

ISBN 978-981-4774-68-0 (Hardcover), 978-1-351-20607-5 (eBook)
www.panstanford.com

and water (H_2O) in 1:23.09:5.25 (v/v %) with 0.01 M ammonium hydroxide (NH_4OH) [6]. The silica sol exhibited hydrophilicity by further condensing or coating on any substrate due to the presence of hydrophilic functional groups on the surface. The silica sol should be modified further with hydrophobic materials to impart hydrophobicity to any substrate. For that, they introduced phenyltriethoxysilane (PTES) as a hydrophobic modifier with the silica sol to improve the surface property from hydrophilic to hydrophobic.

Similarly, silica-based hydrophobic organic-inorganic hybrid materials were also synthesized by mixing TEOS and (heptadecafluoro-1,1,2,2-tetrahydrodecyl) triethoxysilane (FOS) precursors, which were hydrolyzed and polycondensed by the addition of ethanol, deionized water and aqueous ammonia solution and stirred for 12 h at room temperature (Fig. 2.1) [7]. Edwie et al. also fabricated hydrophobic-hydrophilic organic-inorganic hybrid membrane by a dry-jet wet phase inversion process. The membrane solution was prepared by dispersing the hydrophobic silica nanoparticles in methanol followed by mixing with *N*-methyl pyrrolidone (NMP) and the addition of polyvinylidene fluoride (PVDF). All the materials were mixed well prior to fabrication of the membrane films. They obtained hollow fibrous membranes by a solvent exchange method in water using the fabricated fibrous membranes, which were freeze-dried in a freeze dryer. The dual layer of hydrophobic-hydrophilic organic-inorganic hybrid hollow porous membranes can be used for desalination applications [7].

The hydrophobic effect of functional organic silane precursors on the fabrication of hydrophobic organic-inorganic hybrid materials was studied by the hydrolysis and polycondensation of TEOS and various silane precursors in the sol-gel method in the presence of ethanol and hydrochloric acid (HCl) [8]. Purcar et al. used various organic functional triethoxysilane (RTES) precursors, such as methyl- (MTES), vinyl- (VTES), phenyl- (PTES), and octyl- (OTES) triethoxysilanes with a TEOS-to-RTES molar ratio of 1:1 [8]. Coating of the above mixtures showed a range of surface hydrophobic properties on a glass substrate. The hydrophobicity of the coating materials depends on the organic chain length of the silane precursors. The authors also reported that the hydrophobicity is due mainly to the carbon chain length present on the silica surface

[8]. A longer carbon chain on the silica surface would show a higher hydrophobic contact angle (CA). The stronger hydrophobicity is also due to the smaller number of polar surface functional groups at the surface, which may reduce the interactions between the hydrophobic coated surfaces and water droplets.

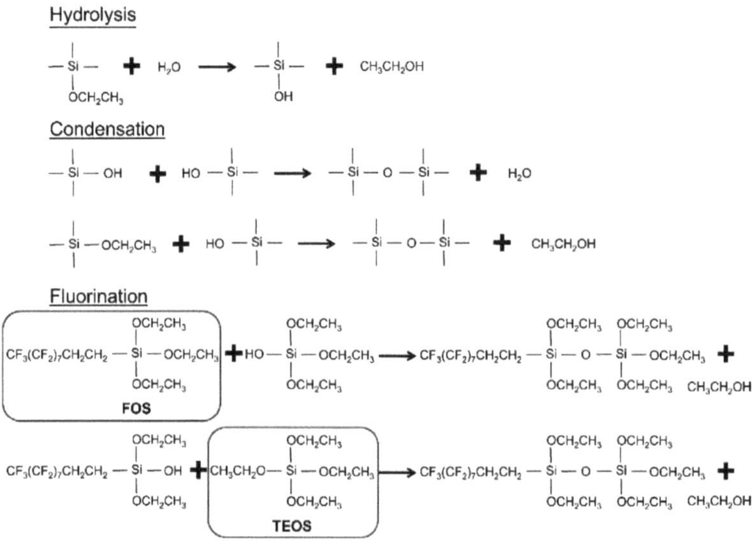

Figure 2.1 Synthesis route of hydrophobic fluorinated silica particles. Reprinted from Ref. [7]. Copyright (2012), with permission from Elsevier.

The hydrophilic and hydrophobic silica nanoparticles were synthesized by a sol-gel method using TEOS and MTMS precursors in two steps. First, the hydrophilic silica nanoparticles were synthesized using TEOS in the presence of methanol, water, and aqueous ammonia solution with a MOS:MeOH:H_2O:NH_4OH molar concentration of 1:15.3:0.71:5 × 10^{-4} [9]. Second, hydrophobic silica nanoparticles were synthesized from an MTMS precursor in the presence of methanol, water, and oxalic acid solution with an MTMS:MeOH:H_2O:oxalic acid molar concentration of 1:16:0.93:0.05.

The silica surface property was also modified using hydrophobic methacrylate and fluoro-silane-based precursors by the sol-gel method [10]. First, they used the general method of synthesis of silica sol by the hydrolysis and condensation of TEOS in isopropanol in the presence of an aqueous HCl catalyst. The surface property of the silica sol was modified by grafting the silica surface using 3-(trimethoxysilyl)propyl methacrylate (TMSPMA) by an in situ method and the silica surface was modified further with 1H,1H,2H,2H-perfluorooctyl triethoxysilane (13F), whereas the fluorosilane acted as a hydrophobic surface enhancer and TMSPMA worked as a coupling agent to bond with plastic and other substrates. The resulting solution showed good adhesion to various substrates and maintained stable hydrophobicity on the substrates due to the presence of hydrophobic low-surface-energy fluorosilane that was also cured easily on the substrate under UV light. Moreover, highly transparent thin films were fabricated by coating the hybrid solution on plastic substrates.

Shi et al. developed novel hydrophobic organic-inorganic solid silica nanospheres and hollow silica nanospheres by forming a micro-emulsion at the beginning by blending a suitable amount of polyoxyethylene nonylphenol ether (NP-7) in cyclohexane, n-butyl alcohol, water, and aqueous ammonia followed by the addition of TEOS and organic silane precursors, such as pheyltriethoxysilane, isobutyltrimethoxysilane, n-octyltrimethoxysilane, dodecyltrimethoxysilane, and propyltrimethoxysilane, respectively [11]. The surface properties of the organic-inorganic hybrid silica nanospheres were tuned carefully by the addition of the above organic silane precursors in the emulsion to improve the hydrophobicity on the silica nanosphere surface. The hydrophobic hollow silica nanospheres were obtained by etching the solid silica nanospheres in an aqueous hydrofluoric acid (HF) solution in the presence of ethanol. The synthesized silica nanospheres and hollow silica nanospheres exhibited strong hydrophobicity.

The hydrophobicity of the synthesized hollow silica nanospheres can be used further for the removal of organic pollutants in water. The authors also analyzed the 4-nonylphenol removal efficiency from water using the synthesized hydrophobic hollow silica nanospheres [11]. The hydrophobic hollow silica nanospheres showed excellent removal efficiency of 4-nonylphenol from water due to the ultrahigh hydrophobicity of the synthesized hollow silica nanospheres. The hydrophobicity of the materials can show higher affinity for the removal of hydrophobic organic compounds in water because of the formation of hydrophobic-hydrophobic interactions between the hydrophobic hollow silica nanospheres and hydrophobic

4-nonylphenol. The presence of hollow silica nanospheres with a large surface area and hydrophobic mesoporous structure may lead to excellent organic compound removal behavior from water.

According to the above concepts, silica precursors are generally used in the sol-gel method for the development of hydrophobic organic-inorganic hybrid coating materials. The synthesis of hydrophobic coating materials was simplified by two ways, such as the coprecursor method and surface derivatization method [12]. The hydrophobic organic-inorganic silica hybrid material was synthesized by the self-hydroxylation and condensation of hydrophilic and hydrophobic silane precursors in the presence of methanol (MeOH) and dual acid (oxalic acid) and base (aq. ammonia) catalysts by a sol-gel method [12]. The hydrophobic coating solution was developed by mixing at TEOS, MeOH, oxalic acid, and NH_4OH at a molar ratio of 1:11.03:0.17:0.58, respectively. A hydrophobic surface was fabricated on a glass substrate by a dip-coating method by simply dipping the glass substrate for a given time in a hydrophobic silica suspension and the substrate was dried at a certain temperature.

Hanna et al. prepared mesoporous silica nanoparticles for the sorption of hydrophobic organic compounds, such as phenol, 3-chlorophenol, 3,5-dichlorophenol, and *p*-toluidine in water [13]. The mesoporous silica nanoparticles were synthesized in two different approaches using TEOS in the presence of water and different surfactants (tetradecyltrimetylammonium bromide [TTAB] and cetyltrimethyl ammonium bromide [CTAB]) and catalysts (aqueous ammonia and sodium hydroxide). The synthesized mesoporous materials displayed excellent sorption of various organic compounds in water. On the other hand, the sorption behavior of organic compounds by the mesoporous materials was found to be related directly to the hydrophobicity of the organic compounds. A thermally stable organic-inorganic hybrid membrane was obtained using TEOS/bis(triethoxysilyl) ethane (BTESE)/MTES mixtures [14]. The sol prepared from the above mixture exhibited good thermal stability due to the presence of thermally stable hydrophobic functional groups at the silica surface.

Hydrophobically modified mesoporous silica-titania organic-inorganic hybrid materials were synthesized by a hydrothermal method using a mixture of BTESE and TEOS with an aqueous solution of octadecyltrimethylammonium chloride (OTMACl) and tetramethyl-ammonium hydroxide (TMAOH) followed by the addition of tetrabutyl orthotitanate [15]. The alcohols present in the solution were removed at 70°C–80°C and transferred to a Teflon bottle. The solution was treated hydrothermally for 3 days at 100°C followed by template removal by a chemical etching method, washing and purification in deionized water and anhydrous ethanol, followed by drying in an oven. Similarly, they also synthesized another hydrophobic hybrid material using MTES instead of BTESE. The synthesized hydrophobic organic-inorganic hybrid materials were used for the catalytic oxidation of cyclohexene with 30% hydrogen peroxide (H_2O_2) at 50°C for 3 h.

Hydrophobic organic-inorganic hybrid nanosilica fillers were synthesized by mixing TEOS and tris-hydroxymethylaminomethane (TRIS) or TEOS and 1,1,1,3,3,3-hexamethyldisilazane (HDMS), followed separately by hydrolysis with the addition of ethanol, water, and acetic acid by a sol-gel method [16]. The resulting sols were gelated and aged to obtain nanosilica particles, which were washed with water and dried at room temperature. Modification of the silica surface with TRIS or HDMS precursors imparted good hydrophobicity to the silica nano particles. The HDMS-modified silica nanoparticles exhibited higher hydrophobicity than TRIS-modified silica nanoparticles due to the formation of a larger particle size and higher degree of nanoparticle agglomeration.

Two types of organic-inorganic hybrid sols were prepared separately by mixing 3-glycidoxypropyl trimethoxysilane (GPTMS) and tetramethoxysilane (TMOS) in aqueous alcohol and 2% nitric acid, and isobutyltrimethoxysilane (IBTMS) in ethanol, water, and acidic acid [17]. The GPTMS/TMOS sol was first coated on a glass substrate by spin coating followed by a second layer coating of IBTMS sol on the same substrate and drying at 120°C in air for 4 h. The resulting substrate showed high transparency and hydrophobicity.

Shimizu et al. developed a transparent and highly insulating hydrophobic organic-inorganic hybrid aerogel by a sol-gel method followed by supercritical drying [18]. The authors prepared several aerogels using various organic silane precursors (ethyltrimethoxysilane [ETMS], vinyltrimethoxysilane [VTMS], MTMS, and TMOS) in the presence of different surfactants (EH-208, polyoxyethylene 2-ethylhexyl ether), CTAB, acid and base catalyst (nitric acid, acetic acid, aqueous tetraethylammonium hydroxide [TEAOH], 35 wt%, and aqueous tetramethylammonium hydroxide [TMAOH, ca. 25%) at different curing temperatures (4°C, 40°C, and 60°C) (Fig. 2.2). The resulting hybrid aerogel exhibited highly hydrophobic, transparent, and excellent insulating properties. On the other hand, the above properties were dependent on the surface reactivity of the organic silane precursors used in fabrication.

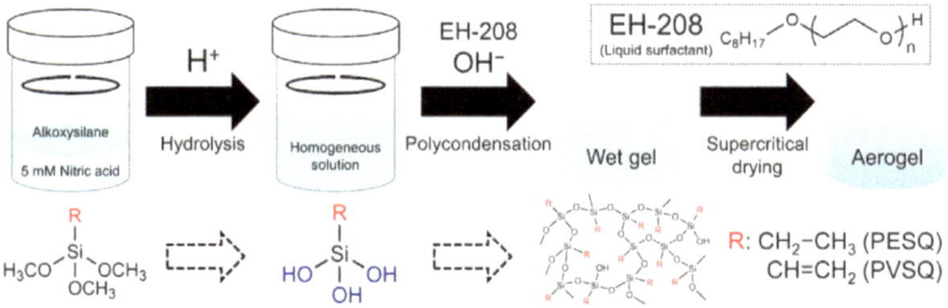

Figure 2.2 Schematic illustration of the synthetic procedure of polyethylsilsesquioxane (PESQ) and polyvinylsilsesquioxane (PVSQ) aerogels. Reprinted from Ref. [18]. Copyright (2016), with permission from the American Chemical Society.

Various polysilsesquioxanes (PSSQs) were synthesized by a sol-gel method and studied in more detail by considering the various factors responsible for the synthesis of PSSQs, including the effects of organic functional silanes, such as TMOS and TEOS responsible for the synthesis of PSSQs by a gel method [19]. Many products have been developed by a reaction of trimethoxysilyl- and triethoxysilyl-substituted monomers with various functional groups (R), such as hydrogen, methyl, ethyl, and vinyl (Table 2.1).

A sol refers to a homogeneous solution. Crystalline products were generally minor contributors to the PSSQ products; most of the product was either in solution or in an oil or resinous phase. Precipitates are insoluble, noncrystalline materials. Gray shading indicates that PSSQ was observed in the indicated physical form. White indicates that the PSSQ failed to afford the indicated physical form.

Loy et al. reported that trimethoxysilyl-substituted monomers reacted faster in some cases (R: methyl, ethyl, and vinyl) than triethoxysilyl-substituted monomers because of the development of an exothermic reaction by the hydrolysis and polycondensation of the materials. The effects of the water stoichiometry, steric, substituent lengths, and other parameters were also examined during the synthesis of organic-inorganic hybrid PSSQs. Moreover, the synthesized organic-inorganic hybrid PSSQs exhibited hydrophobicity based on the surface functional groups present at the PSSQs.

Table 2.1 Range of products obtained from organotrimethoxysilanes, $RSi(OMe)_3$

R Group	Gel	Sol	Crystal	Oil	Resin	Precip.
H						
Methyl						
Ethyl						
Propyl						
n-Butyl						
i-Butyl						
t-Butyl						
Hexyl						
Octyl						
Decyl						
Dodecyl						
Hexadecyl						
Octadecyl						
Cyclohexyl						
Vinyl						
Phenyl						
Phenethyl						
Chloromethyl						
(p-Cloromethyl) phenyl						
Tridecafluoro-1,1,2,2-tetrahydrooctyl						

Source: Adapted from Ref. [19]. Copyright (2000), with permission from the American Chemical Society.

Nakanishi et al. also developed hydrophobic organic-inorganic PSSQ monoliths using a range of alkyltrialkoxysilanes, such as TMOS, MTMS, VTMS, and allyltrimethoxysilane (ATMS), and alkylene-bridged alkoxysilanes, such as 1,1-bis(trimethoxysilyl) methane (BTMM), 1,2-bis(trimethoxysilyl)ethane (BTME), 1,3-bis (trimethoxysilyl)propane (BTMP), and 1,6-bis(trimethoxysilyl) hexane (BTMH) [20]. The resulting hybrids had controlled macro- and mesoporous structures with a hydrophobic surface property.

2.2.2 Emulsion Synthesis

In general, the emulsion method involves the use of a surfactant to stabilize the materials in an oil–water medium. Moreover, the presence of a surfactant allows control of the particle size, shape, aggregation behavior, and the deposition of hydrophilic or hydrophobic materials. Recently, hydrophobic organic-inorganic nanohybrids synthesized by an emulsion method have attracted considerable interest for applications as coating materials on a wide variety of substrates because of the simple and inexpensive synthesis, easy usage, controlled size and shape, and excellent storage capability of the hydrophobic organic-inorganic hybrid materials by emulsion synthesis under wide environmental conditions.

López et al. synthesized highly hydrophobic organic-inorganic hybrid waterborne coatings by the mini-emulsion polymerization of 1H,1H,2H,2H-perfluorodecyl acrylate (PFDA) and 2-ethylhexyl acrylate (2EHA) using an anionic surfactant (Dowfax 2A1 (alkyl diphenyloxide disulfonate)) and azobisisobutyronitrile (AIBN) initiator [21, 22]. The aqueous phase of the solution was prepared by dissolving the monomer in AIBN, whereas the organic phase was prepared using the surfactant. Mini-emulsion polymerization was carried out by the slow addition of an aqueous phase to the organic phase in a nitrogen atmosphere at 70°C. The latex film casted on a glass substrate showed hydrophobicity (130° ± 6°) (Fig. 2.3) due to the presence of highly hydrophobic fluorine chain in the PFDA polymer, which forms stable and hard particles on the casted substrate surface. Moreover, the surface property can be tuned by increasing or decreasing the concentrations of PFDA and 2EHA monomers in the mini-emulsion polymerization process.

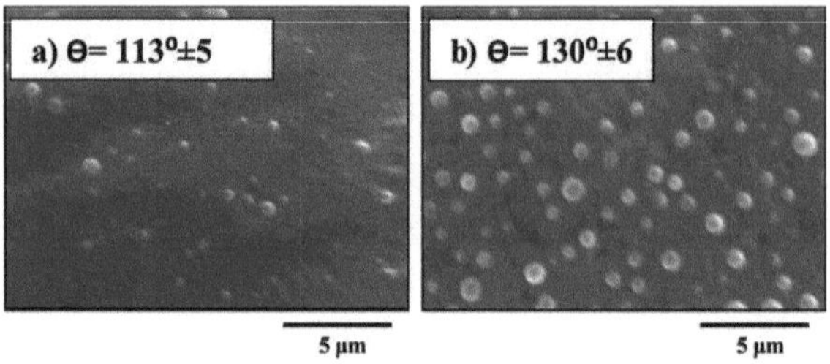

Figure 2.3 Effect of drying upside down on PFDA distribution and contact angle for the 90 μm wet films of blends of latex 1 and 4: (a) Latex1/latex 4 = 20/80 wt/wt; (b) latex1/latex 4 = 50/50 wt/wt. Reprinted from Ref. [22]. Copyright (2016), with permission from the American Chemical Society.

An acrylic-based latex resin with a controlled particle size and core–shell structure containing fluorine functional groups was developed by multistage emulsion polymerization [23]. The inhibitor in all the monomers was removed by passing it through the column before being used for polymerization. The acrylate monomers, such as butyl acrylate (BA), methyl methacrylate (MMA), and 3-(trimethoxysilyl)propyl methacrylate (TMSPMA) with a composition of 49:43:8 by weight, were injected into a round-bottom flask (RBF) condensed with water control and kept in a nitrogen atmosphere in an oil bath at 70°C. The SDS surfactant and KPS initiator were added to the solution and subject to polymerization.

The pH of the colloidal latex obtained by emulsion polymerization was adjusted to 7 using sodium hydrogen carbonate ($NaHCO_3$), and sodium dihydrogen phosphate (NaH_2PO_4) solutions to prevent the hydrolysis and condensation of TMSPMA. The particles size in the acrylate latex was controlled by the further addition of a mixture of anionic/nonionic surfactants (KPS and mixed SDS/Brij58P), while the latex was copolymerized further on the shell with fluorine-based monomers, such as 2,2,2-trifluoroethyl methacrylate (TFEMA) or 3,3,4,4,5,5,6,6,7,7,8,8,8-heptadecafluorodecyl methacrylate (FMA) in the presence of different cyclodextrins as phase transfer catalysts. The organic-inorganic hybrid latex synthesized using this method exhibited excellent hydrophobicity by fabricating on a glass substrate. The surface hydrophobicity was increased by adding the fluorine-based compounds.

Highly monodispersed crosslinked polystyrene (PS) spheres were also synthesized by the emulsion polymerization of styrene monomer and the surface properties were modified using polydimethylsiloxane (PDMS) [24]. The modified hybrid PS/PDMS spheres exhibited stable surface hydrophobicity. Moreover, the surface hydrophobicity can be tuned by changing the concentration of PDMS, where different colors are observed on the coated substrates.

2.2.3 Hydro- and Solvothermal Methods

The hydrothermal and solvothermal method are used generally for the synthesis of micro- or nanoparticles with controlled size or the fabrication of a controlled surface morphology by the deposition of uniform micro- or nanoparticles on a substrate. Water or organic solvents were used as a dispersing agent for the hydrothermal or solvothermal method. The particle size was controlled by the presence of a surfactant or structure-directing agent (SDA) and by the self-assembly of the particles. Huang et al. synthesized a novel poly(3,4-ethylenedioxythiophene) (PEDOT):poly(4-styrenesulfonate) (PSS) by emulsion synthesis followed by hydrothermal modification of the PEDOT:PSS film by the seed growth of zinc oxide (ZnO) nanoparticles [25]. In general, PEDOT:PSS can show excellent conductivity and transparency on a coated substrate. On the other hand, the hybrid organic-inorganic nanohybrids developed by the deposition and growth of ZnO rods on a PEDOT:PSS film can enhance the weather stability, hydrophobicity, pH buffering ability, and acid/alkali resistance due to the excellent property of the ZnO nanoparticles, which form a uniform deposition on the substrate surface.

Myint et al. also reported the hydrophobicity and switchable surface property under UV/IR irradiation by the surface coverage of ZnO microrods [26]. Initially, ZnO nanocrystallites were prepared by a coprecipitation method using zinc acetate as the starting material followed by the dropwise addition of sodium hydroxide in an isopropanol solution. The material was then hydrolyzed at 60°C for 2 h to form the ZnO nanocrystallites, and annealed at 250°C in air for the removal of impurities. The resulting ZnO nanocrystallites were used further for the growth of ZnO microrods. The ZnO nanocrystallites deposited on a glass substrate imparted hydrophilicity. In contrast, the ZnO microrods were grown on a glass substrate by hydrothermal growth in a chemical bath using zinc nitrate hexahydrate and hexamethylenetetramine (90°C), washed with deionized water and annealed at 250°C in air. The fabricated substrate was hydrophobic and showed excellent UV and IR irradiation due to the deposition of a thin layer and the aggregation behavior of ZnO (Fig. 2.4). The hydrophobicity of the substrate surface can be switched to hydrophilic under UV and IR irradiation with respect to the irradiation time and regain the original surface properties by the absence of UV/IR irradiation.

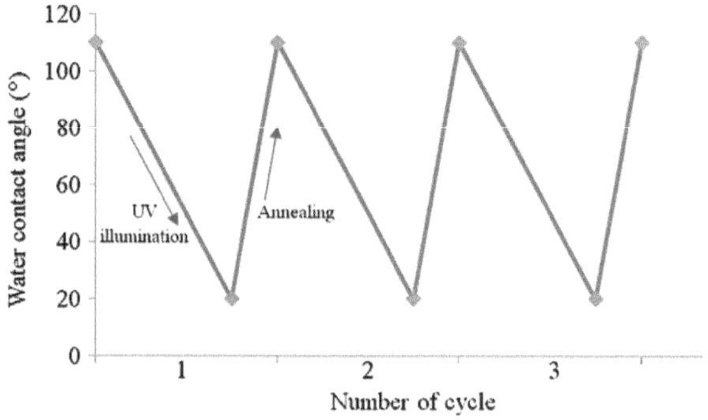

Figure 2.4 Reversible wettability conversion (hydrophobic–hydrophilic) of ZnO microrod–coated glass substrate under UV illumination (1.0 mW/cm^2) and annealing (heating) in the ambient. Reprinted from Ref. [26]. Copyright (2013), with permission from Elsevier.

Hydrotalcite/hydromagnesite deposition was carried out on precleaned Mg alloy (AZ31) sheets substrate by a hydrothermal method [27]. The hydrotalcite/hydromagnesite solution was prepared by dissolving magnesium nitrate hexahydrate ($Mg(NO_3)_2 \cdot 6H_2O$) and aluminum nitrate nonahydrate ($Al(NO_3)_3 \cdot 9H_2O$) in deionized water, which were then mixed slowly with a sodium carbonate and ammonia solution and kept at 333 K with stirring in a three-neck round-bottom flask. The solution was transferred to a Teflon-lined autoclave and the Mg alloy (AZ31) sheets substrate kept horizontally and heated to 398 K for 12 h to deposit the hydrotalcite/hydromagnesite on the substrate. The hydrotalcite/hydromagnesite deposited substrate was cleaned in deionized water followed by drying and surface-hydrophobized with bis-(3-triethoxysilypropyl) tetrasulfide (silane-coupling agent) (Fig. 2.5). Zhang et al. also synthesized mesoporous nanoparticles (MNPs) with a controlled particle size by a solvothermal method and modified the surface property of the amino silane precursor [28]. The magnetic-amino silane surface was then capped with β-cyclodextrin (β-CD), which was called M-CD (Scheme 2.1). The addition of M-CD nanoparticles to the oil–water surface can attach to the surface and alter the surface property from hydrophilic to hydrophobic via the host–guest inclusion complexes (ICs) mechanism. The magnetic nanoparticle inclusion complexes (M-ICs) can also display demulsification properties for various types of emulsions.

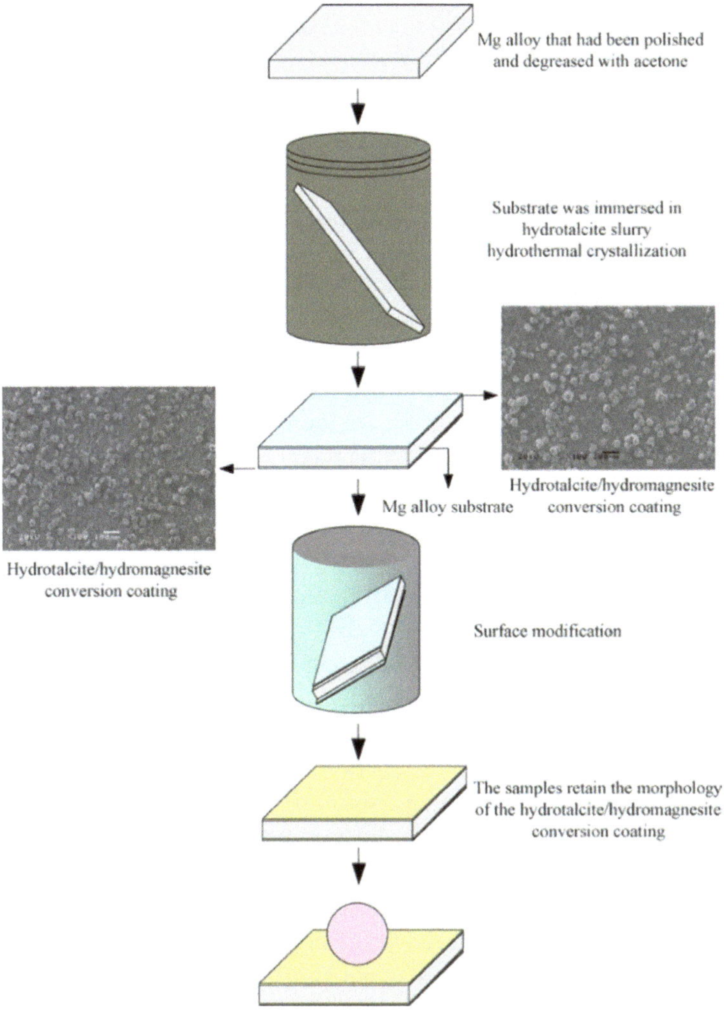

Figure 2.5 Illustration of the fabricated hydrotalcite/hydromagnesite film by in situ hydrothermal crystallization on Mg alloy substrate and surface modification in order to make hydrophobic surface. Reprinted from Ref. [27]. Copyright (2010), with permission from Elsevier.

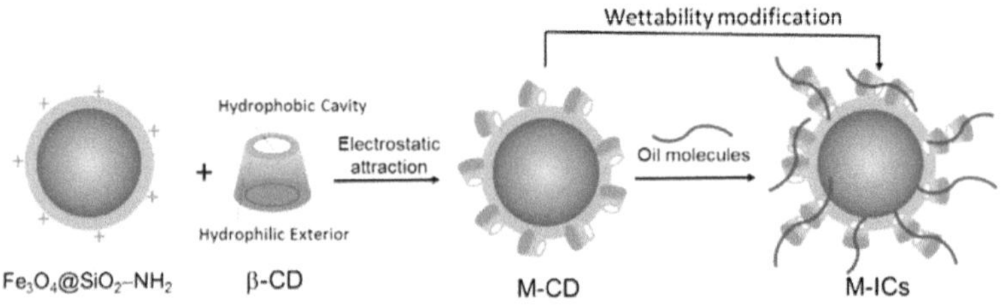

Scheme 2.1 Schematic illustration of the interaction between MNPs and β-CD and the wettability modification of M-CD due to the formation of M-ICs. Reprinted from Ref. [28]. Copyright (2016), with permission from the American Chemical Society.

2.2.4 Surface Grafting and Modifications

Surface grafting or modification is a simple method of modifying the surface property of a material using various grafting agents or crosslinkers. Surface grafting is used widely for the fabrication of hydrophobic organic-inorganic hybrid materials. The main advantages of surface grafting are their simplicity and easy surface modification with various functional groups for the desired applications.

Nagappan et al. synthesized a hydrophobic organic-inorganic hybrid material by surface grafting under the hydrosilylation of polymethylhydrosiloxane (PMHS) and 2,2,3,4,4,4-hexafluorobutyl methacrylate (HFBMA) in the presence of a platinum divinyl tetramethyldisiloxane complex in a vinyl-terminated polydimethylsiloxane catalyst under a nitrogen atmosphere [29]. The fluoro-surface-grafted polymethylsiloxane (FPMS) hybrid material was modified further by a reaction with TEOS and hydrolyzed in the presence of ethanol and water (Scheme 2.2) [29]. A highly transparent organic-inorganic hybrid material was also synthesized using polyvinyl chloride (PVC), ferric chloride hexahydrate ($FeCl_3 \cdot 6H_2O$), and trimethylolpropane tris(3-mercaptopropionate) (TMSH). The hybrid material (PVCFeS) was synthesized in two steps by first preparing an iron-based precursor, followed by surface grafting the precursor with PVC (Scheme 2.3) [30]. The resulting hybrid material solution showed transparent and excellent stability after storage for long periods of time.

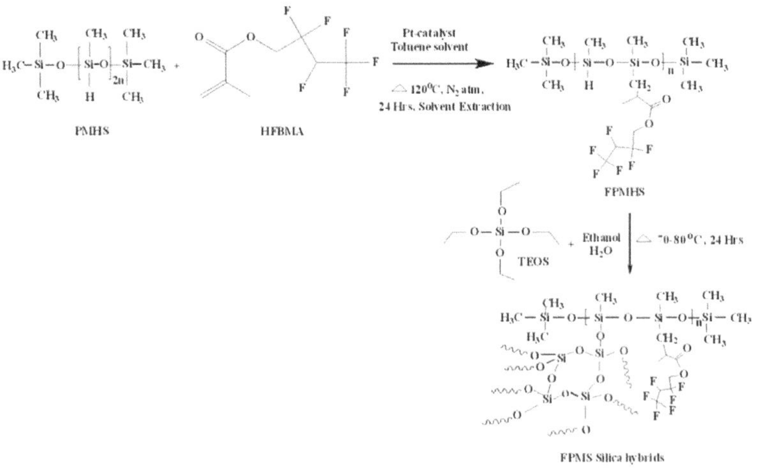

Scheme 2.2 Synthesis of the FPMS/silica hybrids. Reprinted from Ref. [29]. Copyright (2013), with permission from Springer.

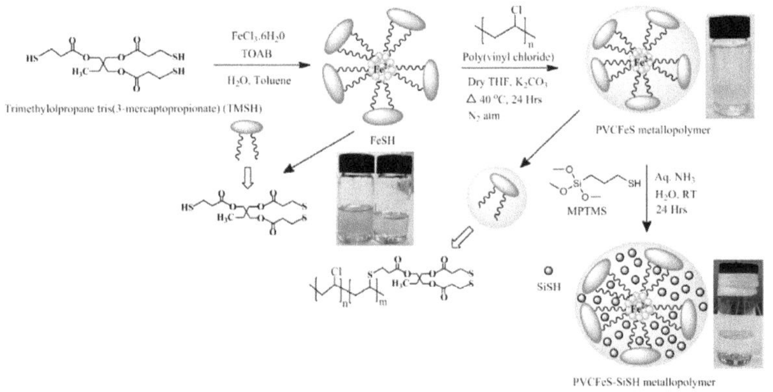

Scheme 2.3 Preparation of the PVCFeS metallopolymer and mercapto silica–functionalized PVCFeS (PCVFeS–SiSH) metallopolymer. Reprinted from Ref. [30] with permission from The Royal Society of Chemistry.

Wang et al. synthesized hydrophobic and thermally stable modified graphene oxide (MGO) by surface grafting the carboxylic functional groups of GO with an amine functional group via covalent bonding [31]. The MGO was synthesized using two approaches: direct amidation of GO with *p*-toluidine using dicyclohexylcarbodiimide (DCC, as a catalyst) or indirect amide formation by modifying the surface property of the carboxylic functional groups in GO using thionyl chloride and further treated with *p*-toluidine. The synthesized MGO showed excellent hydrophobicity after being coated on a substrate due to the presence of a hydrophobic methyl group at the *p*-toluidine moiety in MGO. Bao et al. synthesized novel surface-modified calcium carbonate ($CaCO_3$) particles using polyvinyl chloride (PVC-OH) macromolecular chains [32].

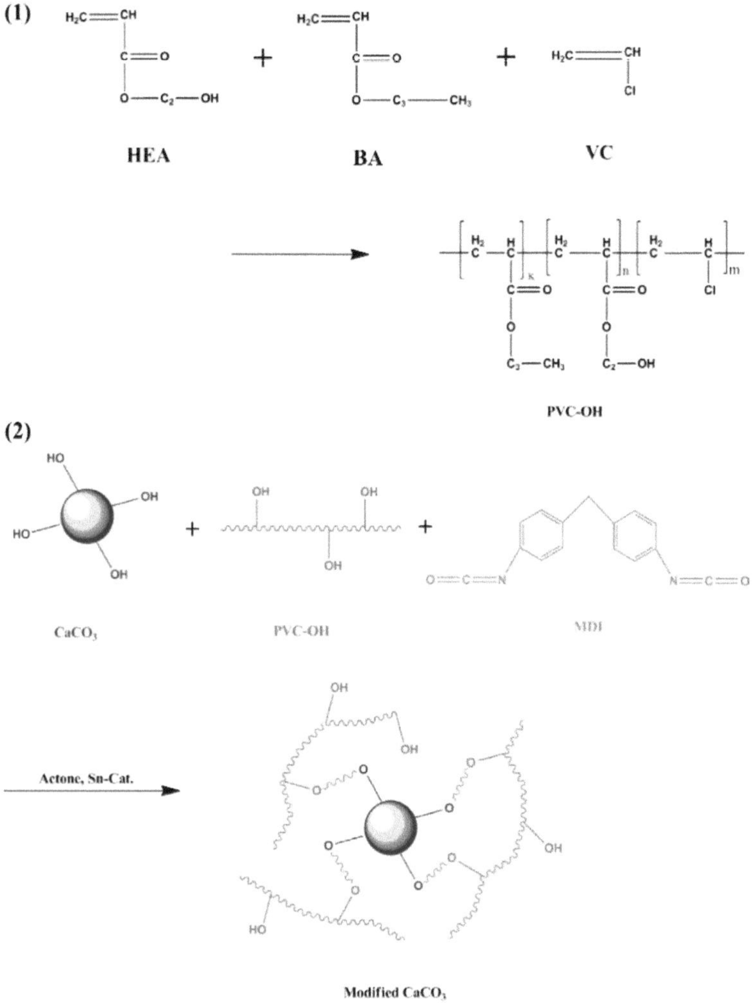

Scheme 2.4 Synthetic route of modified $CaCO_3$ particles. Reprinted from Ref. [32]. Copyright (2015), with permission from Elsevier.

The PVC-OH macromolecular chains were synthesized by the copolymerization of vinyl chloride (VC) monomer with hydroxyethyl acrylate (HEA) and butyl acrylate (BA) monomers with ammonium hydrogen carbonate (NH_4HCO_3), deionized water, dibutyltin dilaurate, KH-20, L-10, tert-butyl peroxyneodecanoate (BNP), and di-(3,5,5-trimethyl hexanoyl) peroxide (TMHP) in a stainless steel reactor under a nitrogen atmosphere. The PVC-OH macromolecular chains exhibited excellent hydrophobicity (CA 125° to 144°) after being coated on a substrate. The surface hydrophobicity was improved partially by further surface grafting with $CaCO_3$ particles using dibutyltin dilaurate and methylene diphenyl diisocyanate (MDI) due to the formation of a hierarchical surface morphology as a result of the formation of micro-nano particles of $CaCO_3$ with PVC-OH (Scheme 2.4).

The organic-inorganic hybrid cotton substrate obtained by surface-treating the natural cellulose fibers with pentafluorobenzoyl chloride in the presence of toluene and pyridine as solvents showed stable hydrophobicity on the cellulose fiber surface (Scheme 2.5) [33]. The fluorine-treated cellulose fiber substrate also exhibited good oleophobicity, chemical inertness, and thermal and oxidative stability due to the presence of low surface energy of a fluorine-based material.

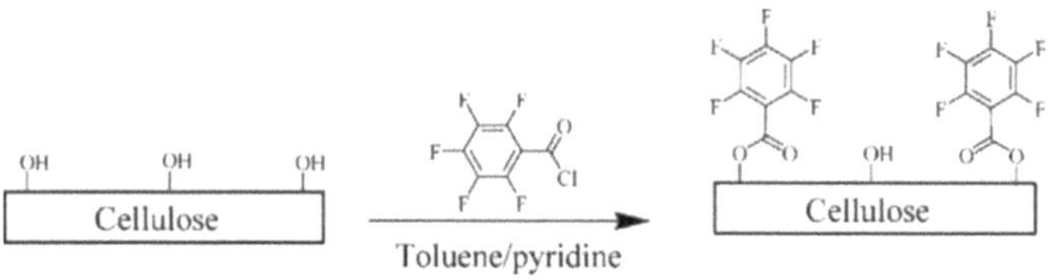

Scheme 2.5 Schematic view of pentafluorobenzoylation of cellulose fibers. Reprinted from Ref. [33]. Copyright (2007), with permission from the American Chemical Society.

An octameric fluorinated polyhedral oligomeric silsesquioxane (F-POSS) was synthesized by the condensation of trialkoxysilanes in alcoholic media using a base catalyst [34]. The synthesized F-POSS showed hydrophobic to superhydrophobic properties based on the fluoroalkyl chain length. Wang et al. also synthesized F-POSS-based hybrid materials by modifying the surface property of the POSS structure with various fluorine chains and further mixed with various concentrations of polymethylmethacrylate (PMMA) [35]. The prepared FPOSS-PMMA hybrid materials exhibited stable hydrophobicity on the coated substrates. The organic-inorganic nanohybrid coating materials were also synthesized and surface-functionalized or grafted with various materials to fabricate the hydrophobic surface on the coated substrates [36–38].

Tapaswi et al. synthesized a novel fluorinated polyimide organic-inorganic hybrid material by surface modification with fluoro-functional groups on the polyimide surface and checked the properties [39]. First, the diamine-based monomers, such as 2,4-diamino-1-fluorobenzene (DAFB) and 4,4'-diamino-2,2'-diiodobiphenyl (DAIB), were synthesized separately and mixed with (hexafluoroisopropylidene) diphthalic anhydride (6FDA) to obtain a poly(amic acid) (PAA) precursor. The PAA was then imidized to obtain the fluorinated polyimide (PI) films (Scheme 2.6). The polyimide films were also obtained in the absence of a fluorine chain at the polyimide surface for comparison. The PI films in the presence of a fluorine chain can show good optical transparency and solubility but inferior mechanical and thermal properties. Furthermore, increasing the concentration of fluorine substituent would improve the surface hydrophobicity and form a stable hydrophobic surface on the film substrates [39].

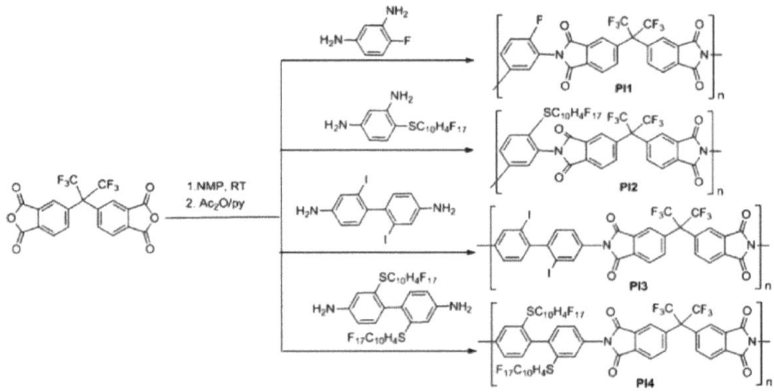

Scheme 2.6 Schematic representation for the synthesis of transparent and hydrophobic fluorinated polyimides. Reprinted from Ref. [39]. Copyright (2014), with permission from WILEY-VCH Verlag GmbH.

2.3 Fabrications of Hydrophobic Organic-Inorganic Nanohybrids

Surface hydrophobicity is an important phenomenon for various surface coating applications due to the water-resisting behavior of the surface. A hydrophobic surface can be obtained by various fabrication methods with the aid of hydrophobic SDAs. Some of the methods involve the fabrication of hydrophobic organic-inorganic hybrid coating surfaces, such as chemical deposition, chemical etching, colloidal assemblies, electrochemistry, layer-by-layer deposition, sol-gel, anodic oxidation, electrospinning, and photolithography. Details of the fabrication of hydrophobic surface properties are given below with more recent and detailed studies.

2.3.1 Spin Coating

Spin coating is a type of coating method used to form a uniform thin film by spreading a hydrophilic or hydrophobic solution or suspension on a range of substrates by a physical force under a constant speed and pressure. The substrate is fixed in a pressure-controlled disc and the solution or suspension is dropped on the substrate and rotated under fixed conditions to deposit a uniform thin film. The spin-coated substrate can exhibit either a homogeneous or heterogeneous rough surface coating based on the selection of materials and other parameters. The surface property of the spin-coated substrates can be altered by varying the parameters, such as the rotation speed, time of rotation, and concentration of the solution or suspension. Spin coating is a widely used fabrication process because the thickness and surface property of the coated substrate can be controlled easily by spin coating.

The hydrophobic-fluorinated polysiloxane-silica hybrid dispersion prepared by the surface grafting of PMHS and HFBMA followed by TEOS modification was spin-coated on a glass substrate at a constant 1000 rpm (rotation per second) for 60 s and annealed at 150°C for 24 h [40]. The fabricated substrate exhibited a smooth to hierarchical rough surface morphology by increasing the concentration of TEOS. Moreover, the fabricated substrate showed transparent and hydrophobic properties on the substrate surface. The surface morphology of the fabricated substrate was also altered by the effect of the amount of ethanol in the hybrid dispersion. Increasing the ethanol concentration to the hydrophobic fluorinated polysiloxane-silica hybrid (FSH) also alters the surface morphology from a hierarchical surface morphology to a smooth surface morphology (Fig. 2.6) [40].

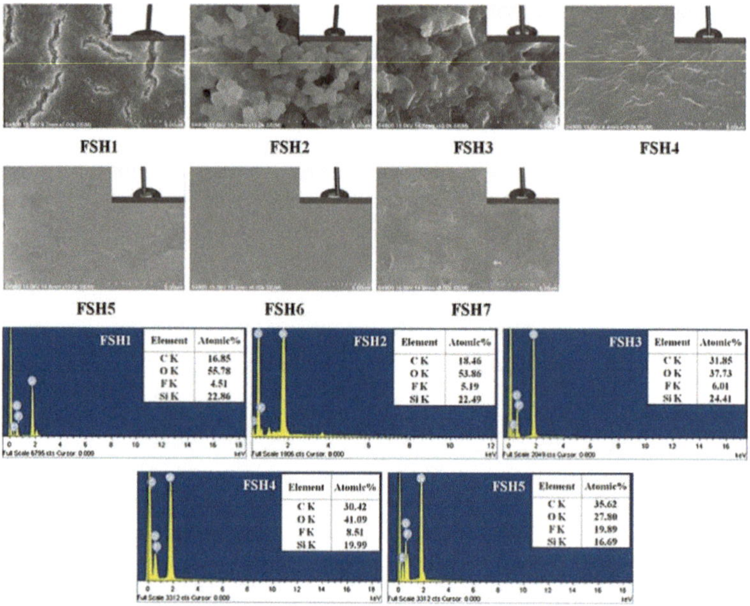

Figure 2.6 High-resolution scanning electron microscopy (HRSEM) images of FSH1–FSH7 hybrids cured at 150°C; the amount of ethanol is increased in the order of FSH1<FSH2<FSH3<FSH4<FSH5<FSH6<FSH7. Insets are the water dynamic contact angle (DCA) drops on the surface of samples for CA measurements. The bottom images illustrate the corresponding scanning electron microscopy/energy dispersive X-ray spectroscopy (SEM-EDS) spectra of FSH1–FSH5 hybrids cured at 150°C. Reprinted from Ref. [40]. Copyright (2013), with permission from Taylor & Francis.

Organic-inorganic functional hybrid films with a 2D nanostructure were fabricated using block copolymer vesicles loaded with inorganic nanoparticles [41]. The hybrid vesicles were spin-coated on a glass substrate to obtain a porous structure. Furthermore, the pure inorganic nanostructures were obtained by plasma-etching the block copolymer template. Using this technique, a well-ordered and uniform porous structure with a honeycomb-like morphology was obtained by the deposition of block copolymer vesicles with similar diameters (Fig. 2.7). Moreover, the surface property of the fabricated surface depends on the thickness and honeycomb structure. The spin-coated substrate obtained using the organic-inorganic hybrid showed hydrophobicity but the hydrophobicity was reduced slightly by the plasma treatment due to the removal of the block copolymer.

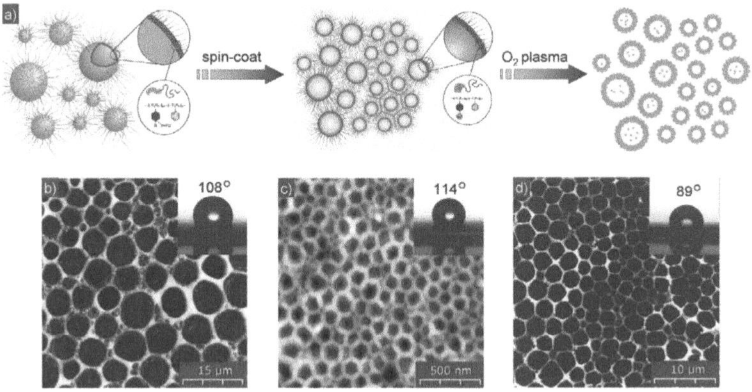

Figure 2.7 (a) Vesicles loaded with metal nanoparticles are prepared in solution, and regular honeycomb structures are fabricated via spin coating. Nanostructured thin films of the inorganic material are obtained after oxygen plasma etching. Depending on the diameter of the vesicles, large (b) or small (c) honeycomb patterns can be fabricated. The structure is preserved during O_2 plasma etching (d). The insets in b–d are optical photographs of water droplets on top of the films, indicating hydrophobic surface properties. Reprinted from Ref. [41]. Copyright (2009), with permission from Elsevier.

The fluorescein-doped silica nanoparticles (FSNPs) were fabricated on a silicon wafer surface using silane-functionalized perfluorophenyl azide (PFPA-silane) by the self-assembly of monolayers of PFPA-silane on the surface followed by the development of polymer arrays and UV irradiation of the substrate [42]. The fabricated substrate exhibited hydrophobicity with excellent protein adsorption behavior. Moreover, the adsorption protein on a silicon wafer surface could be detected easily by labeling with fluorescein. Similarly, a functional poly(allylamine hydrochloride) (PAAm)-PFPA was fabricated on the silicon wafer surface by developing a carbohydrate microarray on the substrate (Scheme 2.7). The PS or polyethylene oxide (PEO) was spin-coated on the substrate followed by UV irradiation and removing the polymer template and carbohydrate by sonication in chloroform and water. Protein adsorption on the fabricated substrate was studied further by labeling the fluorescein material (Scheme 2.7).

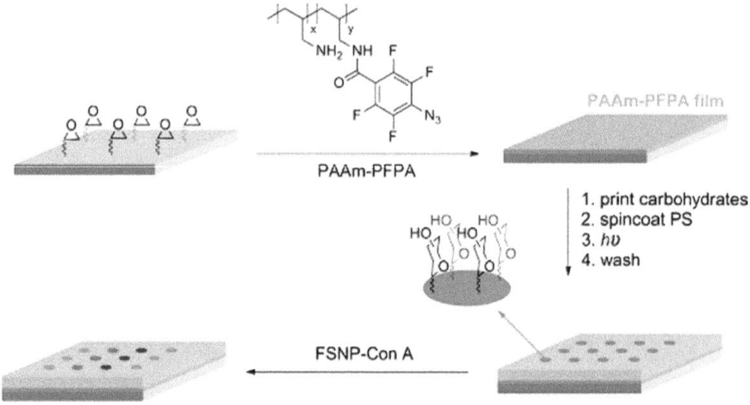

Scheme 2.7 Fabrication of carbohydrate microarrays followed by treating with fluorescein-doped silica nanoparticles (FSNP)-Concanavalin A (Con A). Reprinted from Ref. [42]. Copyright (2013), with permission from the American Chemical Society.

A sulfonated polystyrene (SPS) ionomer solution synthesized from PS and lithium (Li^+) sulfonate ionomers (LiSPS) and spin-coated on a silicone wafer substrate also showed stable hydrophobicity on the surface [43]. In contrast, Shahimin et al. fabricated an elastomeric polydimethylsiloxane (PDMS)–multiwalled carbon nanotube (MWCNT) hybrid surface on a glass substrate [44]. The hybrid PDMS-MWCNT solution prepared at various concentrations was spin-coated on the glass substrate and dried at 80°C for 30 min. The fabricated hybrid substrate exhibited hydrophobicity according to the PDMS-MWCNT concentration.

2.3.2 Dip Coating

Dip coating is a widely used method for the fabrication of hydrophobic organic-inorganic hybrid surfaces on a substrate because of the simple fabrication approach of the organic-inorganic hybrid material on any substrate by simple dipping and followed by subsequent washing and drying of the substrate. The simple approach of a dipping method also reduces the material consumption and cost of the materials. Furthermore, dip coating also allows the uniform deposition of the hybrid coating material on the substrate. Hydrophobic silica and titania nanoparticles were fabricated on a glass substrate by a sol-gel method using titanium isopropoxide (TIP) and MTMS precursors and dip-coated on the precleaned aluminum metal surfaces [45]. The coated substrate exhibited hydrophobic and photoactive properties due to the formation of a hierarchical surface morphology.

Moreover, the hydrophobic hybrid substrate also exhibited the photocatalytic degradation of organic dyes, such as methylene blue (MB). On the other hand, the PMMA/SiO_2/TiO_2 organic-inorganic hybrid composite thin films fabricated by dip coating into the solution obtained by a sol-gel approach of mixing the solution of TMOS to the solution of TMSPMA and MMA in the presence of water, HCl, and tetrabutyl titanate (TBT) and surface treated with perfluoroalkylsulfonyl alkyl triakoxy silane (FC-922) can also exhibit stable hydrophobicity on the coated substrate [46].

Shi et al. fabricated polytetrafluoroethylene (PTFE)@ polypyrrole(PPy)/nanosilver (Ag) composite membranes by immersing the (PPy)/nanosilver (Ag) composite solution in the PTFE membranes that showed excellent hydrophobic and antibacterial properties [47]. The copper nanowall arrays fabricated on the zinc foil substrate by dip coating in a $CuSO_4$ solution also exhibited hydrophobicity [48]. The effects of the surface morphology and surface wettability was checked at various dipping times. A flower-like micropillar and dense nanowall array surface morphology with a double hierarchical structure was developed on the fabricated substrate surface, which may be responsible for the surface hydrophobicity. Moreover, the surface hydrophobicity also depends on the coating time in the $CuSO_4$ solution.

A simple hydrophobic surface was also fabricated by dip coating into the solution made by the hydrosilylation of 1H,1H,2H,2H-perfluorooctyl (PFD) or octafluoropentyloxypropyl (OFP) with trimethoxysilylethyl (TMS) followed by drying at 120°C (Fig. 2.8) [49]. The formation of stable surface hydrophobicity on the coated substrate is due to the presence of low surface fluorine-based materials. The Mg-Al-layered double-hydroxide films and semiconducting surface-anchored metal-organic framework

(SURMOF) film surfaces also exhibited excellent hydrophobicity by a dip-coating method [50, 51].

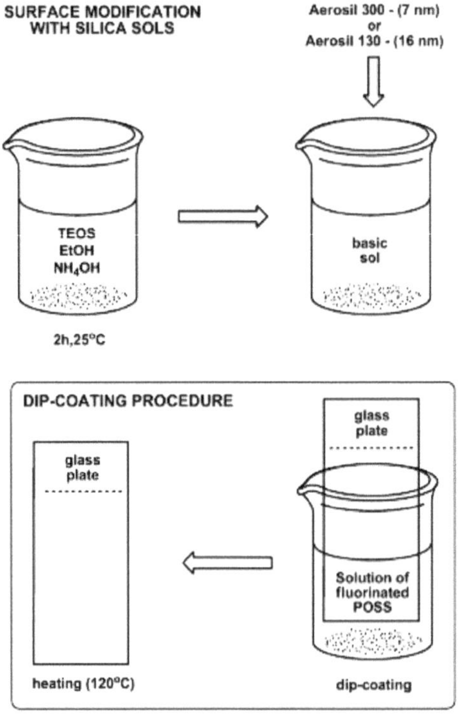

Figure 2.8 General procedure for the modification of glass plates. Reprinted from Ref. [49] with permission from Springer.

2.3.3 Spray Coating

Spray coating is the simplest means of depositing uniform coatings on any substrate and is the most widely used method for a commercial product. Moreover, spray coating may have several advantages than other types of coatings. Spray coating can reduce the coating time and provide coating materials, as well as allow the deposition of uniform coatings on large surface areas of any substrate, improving the transparency and decreasing the curing time on the coated substrate. Spray coating can be conducted either by manually or automatically with a controlled pressure, nozzle diameter, and electricity. The electrostatic spraying approach is used widely in industrial product development because of the controlled deposition of charged particles on the conducting substrate and that it can be cured readily in a shorter time. In contrast, the pressure-controlled spray coating approach is used widely in the development of laboratory coating materials. This is due to the simple approach of forming a more uniform coating on a wide variety of substrates.

A hydrophobic reduced graphene oxide (RGO) paper was fabricated by the spray coating of graphene oxide (GO) followed by the reduction of GO by a thermal treatment [52]. GO is hydrophilic due to the presence of a large number of carboxyl, epoxy, and hydroxyl functional groups at the surface, whereas the hydrophilicity is partially changed to hydrophobic by the reduction of hydrophilic functional groups in GO under thermal treatment. Water adhesion on the RGO paper surface is reduced by increasing the thermal treatment temperature. This simple approach of fabricating a hydrophobic paper substrate with the help of RGO can be used for the separation of oil–water mixtures.

The organic-inorganic hybrid material was obtained by modifying the surface properties of titanium dioxide (TiO_2) nanoparticles using perfluoroalkylsilane (PFAS) [53]. The TiO_2 nanoparticles were dispersed in ethanol and distilled water followed by the slow addition of PFAS under ultrasonication for 4 h in the dark at room temperature (Fig. 2.9). The PFAS-modified TiO_2 nanoparticles were washed with ethanol and distilled water and dried at 65°C for 6 h in a vacuum. The hydrophobic TiO_2 nanoparticles were modified further with PTFE micropowders and fluoroethylene vinyl ether (FEVE). The hydrophobically modified TiO_2 nanoparticles exhibited excellent photocatalytic activity.

Milionis et al. fabricated a novel micro-nano hierarchical patterned surface morphology by first developing a patterned micropillar surface morphology by spin-coating SU-8 on a silicon wafer surface followed by soft baking and UV curing (Fig. 2.10) [54]. The fabricated micropillars were modified further by spraying a uniform deposition of PTFE (micrometer) particle dispersion on the micropillars followed by the deposition of an iron oxide layer and then by a PTFE particle layer and dried overnight. The fabricated substrate exhibited excellent hydrophobicity on a flat surface. This became superhydrophobic on a patterned surface as well as by increasing the surface roughness on the patterned surface. Moreover, the hydrophobicity and superhydrophobicity were maintained with stable surface properties under harsh environmental conditions.

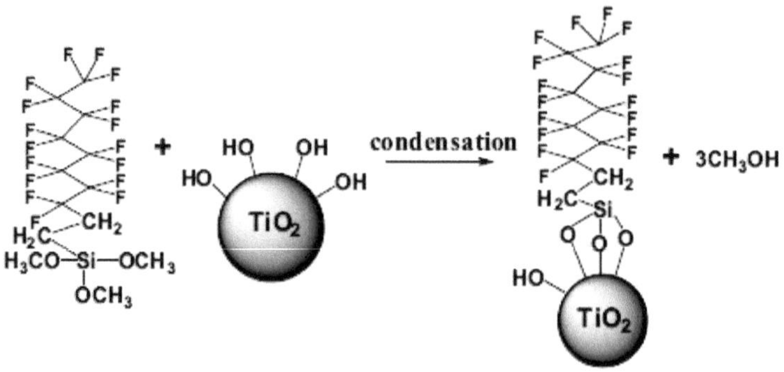

Figure 2.9 Mechanism of surface modification of TiO_2. Reprinted from Ref. [53]. Copyright (2015), with permission from Elsevier.

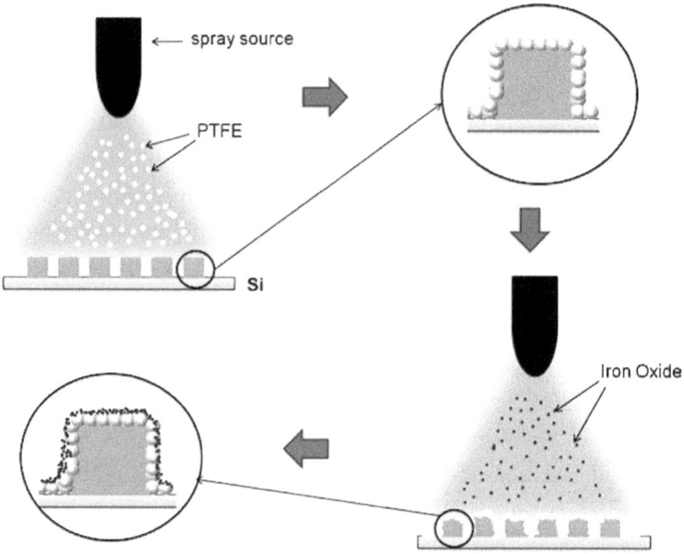

Figure 2.10 Sketch illustrating the general strategy followed for the preparation of the samples. The spraying onto the SU-8 micropillars was performed in two steps. Initially, PTFE submicrometer particles 3 wt% in acetone (top panel) were used. At the second step, a colloidal iron oxide NP dispersion 0.06 wt% in chloroform (bottom panel) was used. At the third step, a second layer of PTFE submicrometer particles is sprayed on top of the NP layer (not shown in the figure). Reprinted from Ref. [54]. Copyright (2012), with permission from Springer.

2.3.4 The SILAR Method

The successive ionic layer adsorption and reaction (SILAR) is also called the SILAR method. The SILAR method is used widely for the deposition of thin films of metal ions and in the fabrication of a uniform thickness and particle size on a substrate. This method is also simple for the fabrication of uniform crystalline thin films, which are used widely in the fabrication of solar cell devices, supercapacitors, and other coating applications. Recently, the SILAR method was also used in the fabrication of a hydrophobic organic-inorganic coating substrate. The method involves a few steps of the successive deposition of anionic and cationic precursors followed by their chemical reaction [55].

First, the substrate was precleaned with suitable solvents and dried to remove the impurities and moisture. The substrate was then immersed in a cationic precursor solution for an optimal time and temperature followed by rinsing the substrate in deionized water. The substrate was immersed further in an anionic precursor solution or catalytic solution for the chemical reaction between cationic and anionic precursors. The substrate was again rinsed with deionized water and dried at a certain temperature and the similar immersion cycles were repeated on the basis of the thickness requirement and the deposition of particles with a uniform size on the substrate.

A crystalline tin sulfide (SnS) thin film was deposited on glass and indium tin oxide (ITO) substrates by the SILAR method using tin(II) sulfate ($SnSO_4$) as the cationic precursor and sodium sulfide (Na_2S) as the anionic precursor [55]. The fabricated crystalline SnS thin film showed excellent transparency and photoluminescence on the substrate. In contrast, Pathan et al. reported the fabrication of TiO_2 thin films on a glass and ITO-coated glass substrates by the SILAR method [56]. Titanium tri-chloride ($TiCl_3$) and NH_4OH were used as the cationic and anionic precursors and the substrate was annealed at 450°C for 6 h to obtain highly crystalline TiO_2 thin films on the substrates [56].

The fabricated TiO_2 thin films exhibited excellent optical and photo-electrochemical properties. Similarly, TiO_2 thin films were also fabricated on glass and fluorine-doped tin oxide (FTO)-coated glass substrates by the SILAR method [57, 58], whereas the TiO_2 thin films deposited on the FTO-coated glass substrate showed hydrophobic

(147°) and superhydrophobic (158°) surface properties based on the aggregation behavior of the TiO_2 thin films on the substrate (Fig. 2.11) [58]. Moreover, the fabricated TiO_2 film electrodes can adsorb rose bengal and methyl violet dyes easily onto the materials, resulting in excellent applicability in the electrode of dye-sensitized solar cells (DSSCs).

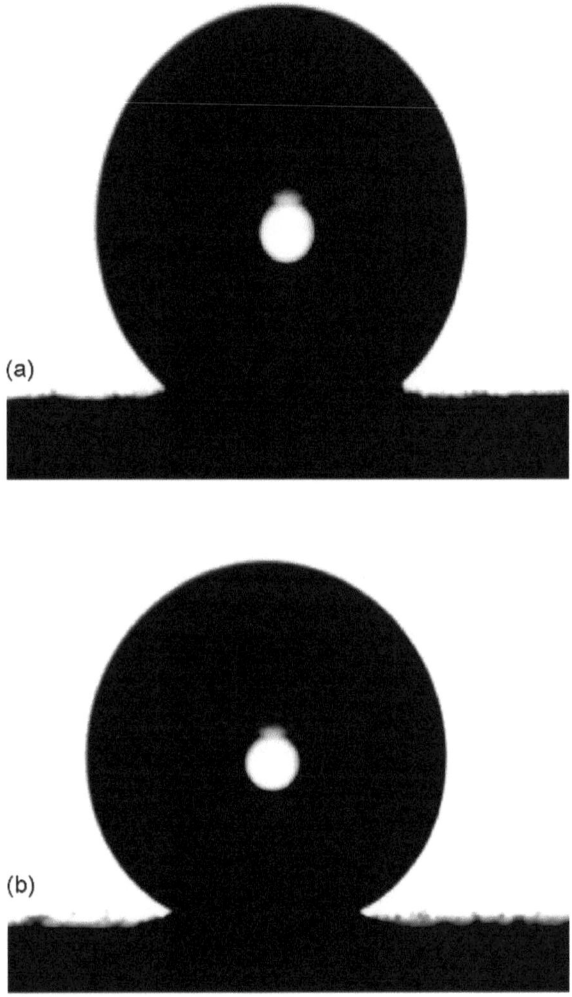

Figure 2.11 Water contact angles (WCAs) of (a) as-deposited (147°) and (b) annealed (158°) TiO_2 films. Reprinted from Ref. [58]. Copyright (2009), with permission from Elsevier.

2.3.5 Electrospinning

Electrospinning is a technique for fabricating nonwoven fabrics with highly porous 3D surface morphologies and a well-ordered uniform nanofibrous mat with a controlled diameter or thickness. The working principle of electrospinning consists mainly of the transportation of continuous filaments from the polymer solution, viscoelastic liquid, or melt liquids using an electrical force [59]. The electrospinning instrumental set-up consists mainly of a few components, such as a high-voltage power supply, a feeding system for holding the liquid solution, a programmable monitor for controlling the spinning system, and a collector with a ground charge (Scheme 2.8).

The nanofibrous diameter and surface morphology were controlled by various parameters, such as the effects of the solvent, solution concentration, pH, applied voltage, distance between the spraying syringe and collector, syringe diameter, time of spraying, and selection of substrates. The nanofibrous mat developed by electrospinning showed many possibilities for use in a variety of applications, such as biomedical applications and drug delivery, separation of heavy metals and oil-in-water or oil–water emulsions, distillation, nanofiltration, supercapacitors, fuel cells, sensors, textiles, and other applications. In most cases, biodegradable and nonbiodegradable polymers are used widely for the preparation of nanofibrous mats.

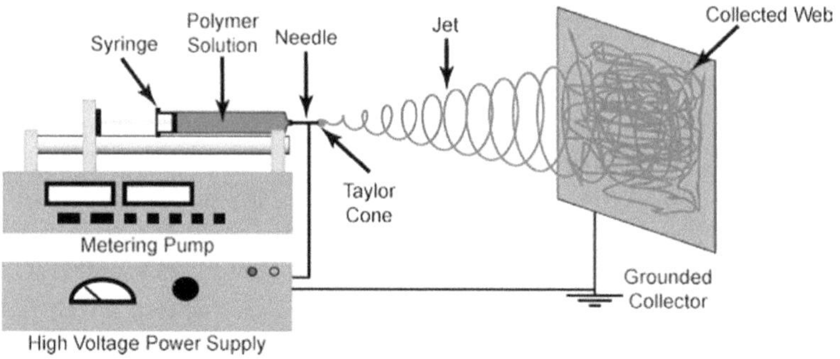

Scheme 2.8 Schematic representation of the basic setup for electrospinning. Reprinted from Ref. [59]. Copyright (2012), with permission from WILEY-VCH Verlag GmbH.

The hydrophobic surface of a nanofibrous mat was obtained by the formation of a hydrophobic nonwoven mat using a hydrophobic polymer solution. Wu et al. fabricated hydrophobic organic-inorganic hybrid nanofibers by first preparing a polyvinyl acetate (PVAc) solution by dissolving PVAc in acetone followed by the dropwise addition of a TiO_2 sol and stirred at room temperature for 24 h [60]. The hybrid nanofibers were obtained from an electrospun PVAc/TiO_2 solution at 15 KV with a syringe diameter of approximately 0.7 mm and a distance between the syringe and aluminum foil of 15 cm.

The fabricated PVAc nanofibrous surface exhibited hydrophobicity with a CA of 109°, whereas the surface CA was reduced from hydrophobic to hydrophilic by increasing the content of TiO_2 nanoparticles in the PVAc solution. On the other hand, the PS nanofibrous membrane fabricated for different electrospinning times exhibited stable hydrophobicity with the CA ranging from 124°–127° [61]. The authors also developed a TiO_2 membrane using an electrospun titanium(IV) butoxide ($Ti(OBu)_4$ solution made with the support of polyvinylpyrrolidone (PVP), toluene, and ethanol. The TiO_2 membrane was hydrophilic, which could be hydrophobized by treatment with a 1H,1H,2H,2H-perfluorooctyltriethoxysilane (POTS) solution in a stainless steel autoclave at 120°C for 1 h.

A hydrophobic organic-inorganic hybrid nanofibrous mat was developed from a mixed hybrid solution of trimethylsilyl cellulose (TMSC) and cellulose triacetate (CTA) in chloroform and electrospun into the continuous nanofibers on an aluminum foil surface [62]. The hybrid nanofibrous surface morphology was controlled by changing the TMSC and CTA concentrations. Moreover, the surface hydrophobicity was also controlled by the TMSC and CTA concentrations.

Polyvinyl butyral (PVB)-based nanofibrous morphology was also developed on a stainless steel wire meshes surface with a minimum applied voltage of 6.80 kV [63]. The pore size of the nanofibrous morphology was controlled using a stainless steel mesh with various pore sizes (Fig. 2.12). The fabricated surface exhibited ultrahigh surface hydrophobicity and excellent separation efficiency in oil–water separation applications. A wide variety of porous nanofibrous mats were also produced using an oil-in-water emulsion electrospinning approach [64]. The oil-in-water emulsion was prepared by dissolving PS in a mixture of chloroform

($CHCl_3$):dimethylformamide (DMF) solvents followed by the addition of a PVP solution prepared in deionized water. The solution was mixed further with fluorescein sodium salt and sorbiton monooleate (SPAN80) surfactant to obtain a stable emulsion. The stable and tunable hydrophobic nanofibrous surface morphology was obtained by electrospinning the emulsion solution under fixed conditions.

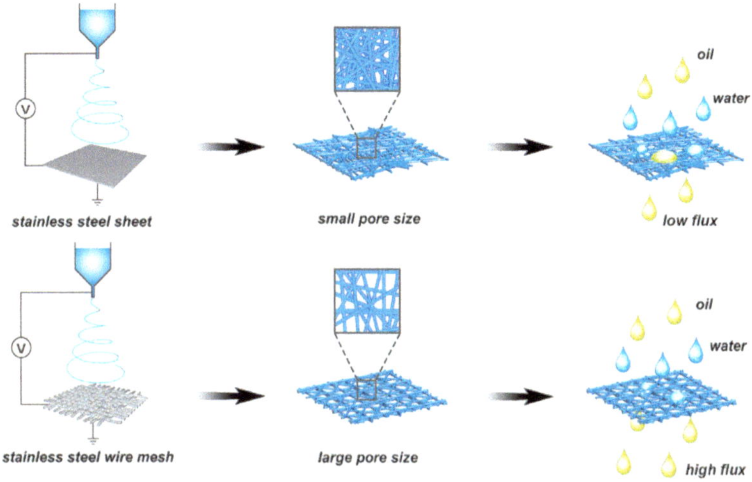

Figure 2.12 Schematic illustration of the process of fabrication of nanofibrous mats with different pore sizes and the corresponding mechanisms for separation of oil–water mixtures. Reprinted from Ref. [63]. Copyright (2016), with permission from the American Chemical Society.

References

1. Kovalchuk, N. M., Trybala, A., Starov, V., Matar, O., and Ivanova, N. (2014). Fluoro- vs hydrocarbon surfactants: why do they differ in wetting performance?, *Adv. Colloid Interface Sci.*, **210**, pp. 65–71.

2. Kuo, S. W., Wu, Y. C., Wang, C. F., and Jeong, K. U. (2009). Preparing low-surface-energy polymer materials by minimizing intermolecular hydrogen-bonding interactions, *J. Phys. Chem. C*, **113**, pp. 20666–20673.

3. Jiang, L., Li, S., Yu, W., Wang, J., Sun, Q., and Li, Z. (2016). Interfacial study on the interaction between hydrophobic nanoparticles and ionic surfactants, *Colloids Surf. A: Physicochem. Eng. Aspects*, **488**, pp. 20–27.

4. Manjumol, K. A., Mini, L., Peer Mohamed A., Hareesh, U. S., and Warrier, K. G. K. (2013). A hybrid sol–gel approach for novel photoactive and hydrophobic titania coatings on aluminium metal surfaces, *RSC Adv.*, **3**, pp. 18062–18070.
5. Mackenzie, J. D., and Bescher, E. (2003). Some factors governing the coating of organic polymers by sol-gel derived hybrid materials, *J. Sol-Gel Sci. Technol.*, **27**, pp. 7–14.
6. Solaree, L. S., Monshi, A., and Ghayour, H. (2015). A new approach for the fabrication of hydrophobic silica coatings on glass using sol–gel method, *Synth. React. Inorg. Met. Org. Nano Metal Chem.*, **45**, pp. 1769–1772.
7. Edwie, F., Teoh, M. M., and Chung, T. S. (2012). Effects of additives on dual-layer hydrophobic–hydrophilic PVDF hollow fiber membranes for membrane distillation and continuous performance, *Chem. Eng. Sci.*, **68**, pp. 567–578.
8. Purcar, V., Stamatin, I., Cinteza, O., Petcu, C., Raditoiu, V., Ghiurea, M., Miclaus, T., and Andronie, A. (2012). Fabrication of hydrophobic and antireflective coatings based on hybrid silica films by sol–gel process, *Surf. Coat. Technol.*, **206**, pp. 4449–4454.
9. Mahadik, D. B., Lee, Y. K., Chavan, N. K., Mahadik, S. A., and Park, H. H. (2016). Monolithic and shrinkage-free hydrophobic silica aerogels via new rapid supercritical extraction process, *J. Supercrit. Fluids*, **107**, pp. 84–91.
10. Chang, C. C., Lin, Z. M., Huang, S. H., and Cheng, L. P. (2015). Preparation of highly transparent 13F-modified nano-silica/polymer hydrophobic hard coatings on plastic substrates, *J. Appl. Sci. Eng.*, **18**, pp. 387–394.
11. Shi, S., Wang, M., Chen, C., Lu, F., Zheng, X., Gao J., and Xu, J. (2013). Preparation of hydrophobic hollow silica nanospheres with porous shells and their application in pollutant removal, *RSC Adv.*, **3**, pp. 1158–1164.
12. Parale, V. G., Mahadik, D. B., Kavale, M. S., Mahadik, S. A., Venkateswara Rao, A., and Mullens, S. (2013). Sol–gel preparation of PTMS modified hydrophobic and transparent silica coatings, *J. Porous Mater.*, **20**, pp. 733–739.
13. Hanna, K., Beurroies, I., Denoyel, R., Desplantier-Giscard, D., Galarneau, A., and Di Renzo, F. (2002). Sorption of hydrophobic molecules by organic/inorganic mesostructures, *J. Colloid Interface Sci.*, **252**, pp. 276–283.
14. Castricum, H. L., Sah, A., Geenevasen, J. A. J., Kreiter, R., Blank, D. H. A., Vente, J. F., and ten Elshof, J. E. (2008). Structure of hybrid

organic-inorganic sols for the preparation of hydrothermally stable membranes, *J. Sol-Gel Sci. Technol.*, **48**, pp. 11–17.

15. Yamamoto, K., Nohara, Y., and Tatsumi, T. (2001). Synthesis and catalysis of Ti-MCM-41 materials with organic–inorganic hybrid network, *Chem. Lett.*, **7**, pp. 648–649.
16. Musante, L., Vázquez, P., Torregrosa-Maciá, R., and Martín-Martínez, J. M. (2010). Synthesis and characterization of new organic–inorganic hybrid nanosilica fillers prepared by sol–gel reaction, *Compos. Interfaces*, **17**, pp. 467–479.
17. Hwang, J. H., Lee, B. I., Klep, V., and Luzinov, I. (2008). Transparent hydrophobic organic–inorganic nanocomposite films, *Mater. Res. Bull.*, **43**, pp. 2652–2657.
18. Shimizu, T., Kanamori, K., Maeno, A., Kaji, H., Doherty, C. M., Falcaro, P., and Nakanishi, K. (2016). Transparent, highly insulating polyethyl- and polyvinylsilsesquioxane aerogels: mechanical improvements by vulcanization for ambient pressure drying, *Chem. Mater.*, **28**, pp. 6860–6868.
19. Loy, D. A., Baugher, B. M., Baugher, C. R., Schneider, D. A., and Rahimian, K. (2000). Substituent effects on the sol-gel chemistry of organotrialkoxysilanes. *Chem. Mater.*, **12**, pp. 3624–3632.
20. Nakanishi, K., and Kanamori, K. (2005). Organic–inorganic hybrid poly(silsesquioxane) monoliths with controlled macro- and mesopores, *J. Mater. Chem.*, **15**, pp. 3776–3786.
21. López, A. B., de la Cal, J. C., and Asua, J. M. (2016). Controlling film topography to form highly hydrophobic waterborne coatings, *Soft Matter*, **12**, pp. 7005–7011.
22. López, A. B. de la Cal J. C., and Asua, J. M. (2016). Highly hydrophobic coatings from waterborne latexes, *Langmuir*, **32**, pp. 7459–7466.
23. Yang, C. C., Castelvetro, V., Bianchi, S., Alderighi, M., and Zhang, Y. (2012). Hierarchical dual-sized film surface morphologies self-generated from fluorinated binary latex blends boost hydrophobicity and lipophobicity, *J. Colloid Interface Sci.*, **378**, pp. 210–221.
24. Tang, B., Zheng, X., Lin, T., and Zhang, S. (2014). Hydrophobic structural color films with bright color and tunable stop-bands, *Dyes Pigm.*, **104**, pp. 146–150.
25. Huang, T. C., Yin, H. E., Chiu, W. Y., and Lee, C. F. (2014). Fabrication of hydrophobic poly(3,4-ethylenedioxythiophene): poly(4-styrenesulfonate)/zinc oxide rod conductive film via hydrothermal method and their performance on weather stability and pH buffering ability, *Thin Solid Films*, **552**, pp. 98–104.

26. Myint, M. T. Z., Kumar, N. S., Hornyak, G. L., and Dutta, J. (2013). Hydrophobic/hydrophilic switching on zinc oxide micro-textured surface, *Appl. Surf. Sci.*, **264**, pp. 344–348.

27. Wang, J., Li, D., Liu, Q., Yin, X., Zhang, Y., Jing, X., and Zhang, M. (2010). Fabrication of hydrophobic surface with hierarchical structure on Mg alloy and its corrosion resistance, *Electrochim. Acta*, **55**, pp. 6897–6906.

28. Zhang, J., Li, Y., Bao, M., Yang, X., and Wang, Z. (2016). Facile fabrication of cyclodextrin-modified magnetic particles for effective demulsification from various types of emulsions, *Environ. Sci. Technol.*, **50**, pp. 8809–8816.

29. Nagappan, S., Choi, M. C., Sung, G., Park, S. S., Moorthy, M. S., Chu, S. W., Lee, W. K., and Ha, C. S. (2013). Highly transparent, hydrophobic fluorinated polymethylsiloxane/silica organic-inorganic hybrids for anti-stain coating, *Macromol. Res.*, **21**, pp. 669–680.

30. Nagappan, S., Park, S. S., Yu, E. J., Cho, J. J., Park, J. J., Lee, W. K., and Ha, C. S. (2013). A highly transparent, amphiphobic, stable and multi-purpose poly(vinyl chloride) metallopolymer for anti-fouling and anti-staining coatings, *J. Mater. Chem. A*, **1**, pp. 12144–12153.

31. Wang, J., Geng, H. Z., Luo, Z. J., Zhang, S., Zhang, J., Liu, J., Yang, H. J., Ma, S., Sun, B., Wang, Y., Da, S. X., and Fua, Y. Q. (2015). Preparation, characterization, and chemical induced hydrophobicity of thermostable amine modified graphene oxide, *RSC Adv.*, **5**, pp. 105393–105399.

32. Bao, L., Yang, S., Luo, X., Lei, J., Cao, Q., and Wang, J. (2015). Fabrication and characterization of a novel hydrophobic $CaCO_3$ grafted by hydroxylated poly(vinyl chloride) chains, *Appl. Surf. Sci.*, **357**, pp. 564–572.

33. Cunha, A. G., Freire, C. S. R., Silvestre, A. J. D., Neto, C. P., Gandini, A., Orblin, E., and Fardim, P. (2007). Highly hydrophobic biopolymers prepared by the surface pentafluorobenzoylation of cellulose substrates, *Biomacromolecules*, **8**, pp. 1347–1352.

34. Mabry, J. M., Vij, A., Iacono, S. T., and Viers, B. D. (2008). Fluorinated polyhedral oligomeric silsesquioxanes, *Angew. Chem. Int. Ed.*, **47**, pp. 4137–4140.

35. Wang, X., Ye, Q., Song, J., Cho, C. M., He, C., and Xu, J. (2015). Fluorinated polyhedral oligomeric silsesquioxanes, *RSC Adv.*, **5**, pp. 4547–4553.

36. Yang, C. H., Pan, Y. W., Guo, J. J., Young, T. H. Chiu, W. Y., and Hsieh, K. H. (2016). Synthesis and application of polyurethane basic organic-inorganic hybrid materials as highly hydrophobic coatings, *J. Polym. Res.*, **23**, p. 29.

37. Lin, C. C., Hsu, S. H., Chang, Y. L., and Su, W. F. (2010). Transparent hydrophobic durable low moisture permeation poly(fluoroimide acrylate)/SiO_2 nanocomposite from solventless photocurable resin system, *J. Mater. Chem.*, **20**, pp. 3084–3091.

38. Zayas, M. S., Gaitor, J. C., Nestor, S. T., Minkowicz, S., Sheng Y., and Mirjafari, A. (2016). Bifunctional hydrophobic ionic liquids: facile synthesis by thiol–ene "click" chemistry, *Green Chem.*, **18**, pp. 2443–2452.

39. Tapaswi, P. K., Choi, M. C., Nagappan, S., and Ha, C. S. (2015). Synthesis and characterization of highly transparent and hydrophobic fluorinated polyimides derived from perfluorodecylthio substituted diamine monomers, *J. Polym. Sci. Part A: Polym. Chem.*, **53**, pp. 479–488.

40. Nagappan, S., Park, J. J., Park, S. S., Hong, S. H., Jeong, Y. S., Lee, W. K., and Ha, C. S. (2013). Polymethylhydrosiloxane based organic-inorganic hybrids for amphiphobic coatings, *Compos. Interfaces*, **20**, pp. 33–44.

41. Fahmi, A., Pietsch, T., Mendoza, C., and Cheval, N. (2009). Functional hybrid materials, *Mater. Today*, **12**, pp. 44–50.

42. Wang, H., Tong, Q., and Yan, M. (2013). Antifouling surfaces for proteins labeled with dye-doped silica nanoparticles, *Anal. Chem.*, **85**, pp. 23–27.

43. Wang, X., and Weiss, R. A. (2012). A facile method for preparing sticky, hydrophobic polymer surfaces, *Langmuir*, **28**, pp. 3298–3305.

44. Shahimin, M. M., Rodzi, N. H. M., Omar, M. F., Poopalan, P., Man, B., and Md Nor, M. N. (2015). Hybrid elastomer-nanotube matrix for hydrophobic surface functionalization, *J. Adhes. Sci. Technol.*, **29**, pp. 532–542.

45. Manjumol, K. A., Mini, L., Mohamed, A. P., Hareesh, U. S., and Warrier, K. G. K. (2013). A hybrid sol–gel approach for novel photoactive and hydrophobic titania coatings on aluminium metal surfaces, *RSC Adv.*, **3**, pp. 18062–18070.

46. Gu, G., Zhang, Z., and Dang, H. (2004). Preparation and characterization of hydrophobic organic–inorganic composite thin films of PMMA/SiO_2/TiO_2 with low friction coefficient, *Appl. Surf. Sci.*, **221**, pp. 129–135.

47. Shi, Z., Zhou, H., Qing, X., Dai, T., and Lu, Y. (2012). Facile fabrication and characterization of poly(tetrafluoroethylene)@polypyrrole/nano-silver composite membranes with conducting and antibacterial property, *Appl. Surf. Sci.*, **258**, pp. 6359–6365.

48. Tian, C. F., Zhang, F. L., Cai, H. H., Fang, J. X., and Ding, B. J. (2013). 'Flower-like' micropillars of copper with porous nanowalls as effective hydrophobic surface, *Mater. Sci. Technol.*, **29**, pp. 446–450.

49. Maciejewski, H., Karasiewicz, J., Dutkiewicz, M., and Marciniec, B. (2014). Hydrophobic materials based on fluorocarbofunctional spherosilicates, *Silicon*, **7**, pp. 201–209.

50. Wang, Y., Zhang, D., and Lu, Z. (2015). Hydrophobic Mg–Al layered double hydroxide film on aluminum: fabrication and microbiologically influenced corrosion resistance properties, *Colloids Surf. A: Physicochem. Eng. Aspects*, **474**, pp. 44–51.

51. Rana, S., Rajendra, R., Dhara, B., Jha, P. K., and Ballav, N. (2016). Highly hydrophobic and chemically rectifiable surface-anchored metal-organic framework thin-film devices, *Adv. Mater. Interfaces*, **3**, p. 1500738.

52. Qin, H., Gong, T., Cho, Y., Shin, C., Lee, C., and Kim, T. (2015). A simple and economical method using grapheme oxide for the fabrication of water/oil separation papers, *RSC Adv.*, **5**, pp. 57860–57864.

53. Zhou, Y., Li, M., Zhong, X., Zhun, Z., Deng, P., and Liu, H. (2015). Hydrophobic composite coatings with photocatalytic self-cleaning properties by micro/nanoparticles mixed with fluorocarbon resin, *Ceram. Int.*, **41**, pp. 5341–5347.

54. Milionis, A., Martiradonna, L., Anyfantis, G. C., Cozzoli, P. D., Bayer, I. S., Fragouli, D., and Athanassiou, A. (2012). Control of the water adhesion on hydrophobic micropillars by spray coating technique, *Colloid Polym. Sci.*, **291**, pp. 401–407.

55. Ghosh, B., Das, M., Banerjee, P., and Das, S. (2008). Fabrication and optical properties of SnS thin films by SILAR method, *Appl. Surf. Sci.*, **254**, pp. 6436–6440.

56. Pathan, H. M., Min, S. K., Desai, J. D., Jung, K. D., and Joo, O. S. (2006). Preparation and characterization of titanium dioxide thin films by SILAR method, *Mater. Chem. Phys.*, **97**, pp. 5–9.

57. Kale, S. S., Mane, R. S., Chung, H., Yoon, M. Y., Lokhande, C. D., and Han, S. H. (2006). Use of successive ionic layer adsorption and reaction (SILAR) method for amorphous titanium dioxide thin films growth, *Appl. Surf. Sci.*, **253**, pp. 421–424.

58. More, A. M., Gunjakar, J. L., Lokhande, C. D., and Joo, O. S. (2009). Fabrication of hydrophobic surface of titanium dioxide films by successive ionic layer adsorption and reaction (SILAR) method, *Appl. Surf. Sci.*, **255**, pp. 6067–6072.

59. Sas, I., Gorga, R. E., Joines, J. A., and Thoney, K. A. (2012). Literature review on superhydrophobic self-cleaning surfaces produced by electrospinning, *J. Polym. Sci. Part B: Polym. Phys.*, **50**, pp. 824–845.

60. Wu, N., Sun, Y., Jiao, Y., and Chen, L. (2013). Effect of inorganic/organic hybrid on the wettability of polymer nanofibrous membranes, *J. Eng. Fibers Fabr.*, **8**, pp. 1–5.

61. Hongzhe, T., and Wu, Z. (2016). Application of percolation theory on hydrophobic membrane with high adhesive force, *Colloids Surf. A: Physicochem. Eng. Aspects*, **498**, pp. 276–282.

62. Taajamaa, L., Kontturi, E., Lainea, J., and Rojas, O. J. (2012). Bicomponent fibre mats with adhesive ultra-hydrophobicity tailored with cellulose derivatives, *J. Mater. Chem.*, **22**, pp. 12072–12082.

63. Song, B., and Xu, Q. (2016). Highly hydrophobic and superoleophilic nanofibrous mats with controllable pore sizes for efficient oil/water separation, *Langmuir*, **32**, pp. 9960–9966.

64. Yazgan, G., Popa, A. M., Rossi, R. M., Maniura-Weber, K., Puigmartí-Luis, J., Crespy, D., and Fortunat, G. (2015). Tunable release of hydrophilic compounds from hydrophobic nanostructured fibers prepared by emulsion electrospinning, *Polymer*, **66**, pp. 268–227.

Chapter 3

Applications of Hydrophobic Organic-Inorganic Nanohybrids

3.1 Introduction

Hydrophobic organic-inorganic nanohybrids exhibit excellent surface hydrophobicity, such as good chemical and thermal stability, mechanical stability, transparency, scratch resistance, and water repellence. The applications of hydrophobic organic-inorganic nanohybrids have wide usability in various fields, such as oil–water separation, catalytic degradation of organic dyes, corrosion resistance, scratch resistance coatings, biomedical applications, and anti-icing. This chapter provides some of the important applications of hydrophobic organic-inorganic nanohybrid materials.

3.2 Applications of Hydrophobic Organic-Inorganic Nanohybrids

3.2.1 Oil Spill Capture and Separation

Oil spills are major environmental pollution events caused by the leakage of oil from oil-developing operations, accidents during the transportation of oil by ships, and other sources of oil leakage in the sea. Oil spills on the water surface can reduce the growth of

Hydrophobic and Superhydrophobic Organic-Inorganic Nanohybrids
Chang-Sik Ha and Saravanan Nagappan

ISBN 978-981-4774-68-0 (Hardcover), 978-1-351-20607-5 (eBook)
www.panstanford.com

plants and cause various health problems to sea creatures, humans, and other mammals. Several natural products have been used for the capture of oil spills on the water surface. The hydrophobic nanofibrous mat developed by Song et al. with variable pore sizes of a stainless steel substrate showed excellent separation efficiency of various oils from oil–water mixtures [1].

The oil–water separation by the hydrophobic organic-inorganic nanohybrids was carried out by tightly fixing the hydrophobic mesh in between two glass tubes. The flux-induced separation of oil and the separation efficiency were calculated based on the amount of oil being separated from the water. The developed hydrophobic nanofibrous mat exhibited excellent separation efficiency (over 99%) for various oils, while maintaining good stability by repeated use for several cycles. The hybrid nanofibrous mat also exhibited excellent separation efficiency to water-in-oil emulsions. The increasing pore size of a nanofibrous mat would also increase the oil flux from a water-in-oil emulsion. The low cost fabrication of a hydrophobic-reduced graphene oxide (RGO) filter also showed excellent oil–water separation behavior [2]. The surface modification of a polyethersulfone (PES) membrane using the polypropylene (PP) technique exhibited excellent filtration properties for various cooking oils being used for frying food [3]. Triglyceride-based compounds are preferential for filtration in a PP-modified PES membrane than other polar compounds.

More recently, magnetic-based hybrid materials have attracted considerable attention in the sorption and separation of various oils from a water surface. The magnetic-responsive behavior of the hydrophobic surface can be controlled, actuated, and separated easily for repeated sorption and separation applications. Pei et al. synthesized hydrophobic magnetic nanoparticles by first synthesizing Fe_3O_4 magnetite nanoparticles by a coprecipitation method followed by a surface treatment of Fe_3O_4 nanoparticles with tetraethoxysilane (TEOS) and methacryloxypropyl trimethoxysilane (MPTMS) [4]. The resulting hydrophobic magnetic nanoparticles showed good oil sorption ability and could be used repeatedly by separating the oil-sorbed magnetic particles using an external magnetic source, washing, and drying for the continuous sorption of oils from the water surface. A hydrophobic magnetic-responsive actuating $MnFe_2O_4$ micromotor was fabricated by surface-treating the synthesized $MnFe_2O_4$ with oleic acid (OA) [5]. The hydrophobic $MnFe_2O_4$@OA exhibited an excellent magnetic response and controlled movement of the tracked and sorbed oil using an external magnet. Highly hydrophobic and magnetic-responsive polymer nanocomposites obtained by a facile low-cost emulsion polymerization approach also exhibited excellent oil sorption and recycling properties from the water surface (Fig. 3.1) [6]. The hydrophobic magnetic nanoparticles were spread over the oil-in-water surface and were actuated magnetically. The spilled oils could be absorbed or adsorbed easily on the surface of the hydrophobic magnetic nanoparticles and recycled easily for several cycles by simple washing and drying of the material.

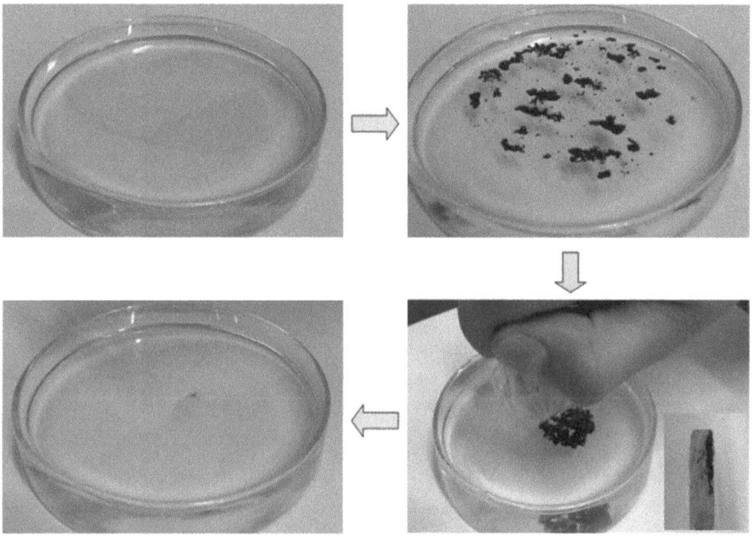

Figure 3.1 Removal of diesel oil from water surface by the as-prepared hydrophobic magnetic nanparticles under magnetic field. The diesel oil was labeled by Sudan I for clarity. Reprinted from Ref. [6]. Copyright (2014), with permission from Elsevier.

A magnetic-responsive and hydrophobic cellulose/TiO_2 aerogel was fabricated by first preparing a cellulose aerogel from a cellulose solution obtained by dissolving cellulose fibers in a sodium hydroxide:thiourea:urea (NTU) aqueous solvent followed by the supercritical CO_2 drying of a cellulose gel [7]. The cellulose aerogel was modified further with magnetic (Fe_3O_4) and TiO_2 nanoparticles.

The fabricated cellulose aerogel exhibited a highly porous structure, hydrophobic and magnetically responsive properties. Moreover, the aerogel exhibited the excellent sorption of oil (28 times of its own weight within 10 min) and the oil-sorbed aerogel could be removed easily from the oil–water surface using an external bar magnet (Fig. 3.2). The sorbed oils were collected by simple squeezing of the oil-sorbed aerogel and reused after washing with ethanol and drying.

Figure 3.2 Photograph of (a) magnetic cellulose/TiO_2 aerogel in a oil–water mixture. (b) The magnetic cellulose/TiO_2 aerogel can be removed by a magnet. (c) The oil layer on water surface was removed. Reprinted from Ref. [7]. Copyright (2014), with permission from Elsevier.

A highly flexible and hydrophobic cellulose aerogel fabricated by Lin et al. also exhibited the excellent sorption of various oils from water surface [8]. Similarly, several hydrophobic, porous, fire-retardant, and stable sponges were also prepared by modifying the sponge surface with a range of materials. The results would lead to the sorption of various oils and organic solvents from the water surface and oil–water mixtures [9–11].

3.2.2 Catalytic Application

Photocatalysis is a technique of accelerating a photochemical material in the presence of a catalyst and a light source. Several inorganic materials show photocatalytic activity under UV light or sunlight. Moreover, inorganic nanomaterials with a controlled size and shape showed excellent photocatalytic activity due to the sorption of the light source by the materials, where, in the presence of vacant electrons, the sorbed electron can be excited under light. This photocatalytic behavior also occurs naturally in plants and tree leaves due to the presence of chlorophyll, which absorbs light energy and transforms carbon dioxide and water to oxygen and glucose by photosynthesis and decomposes them to carbon dioxide and water by photocatalysis. Titanium dioxide (TiO_2) and zinc oxide (ZnO), as well as other metal particles also possess photocatalytic activity.

The titanium-modified mesoporous silica nanoparticles (Ti-MCM-41) and hydrophobically modified Ti-MCM-41 organic-inorganic hybrid materials synthesized from organoalkoxysilanes, such as bis(triethoxysilyl) ethane (BTESE) and methyltriethoxysilane (MTES), by Yamamoto et al. exhibited excellent catalytic activity for the oxidation of cyclohexene (olefin) using hydrogen peroxide (H_2O_2) as the oxidant [12]. The hydrophobically modified Ti-MCM-41 hybrid materials exhibited better catalytic activity and selectivity than the unmodified Ti-MCM-41 and MCM-41. The higher hydrophobicity of the hydrophobically modified Ti-MCM-41 decreased the H_2O_2:olefin concentration ratio and performed a better catalytic reaction.

Poly(methacrylate) (PMA)-modified nanoporous silica organic-inorganic hybrid materials immobilized with amino lipase polystyrene (PS) from the *Pseudomonas cepacia* (PCL, 30 U mg^{-1}) enzyme exhibited excellent hydrophobicity and catalytic activity between (R,S)-1-phenylethanol and vinyl acetate due to the hydrophobic nature and crowded atmosphere in the surrounding immobilized enzyme, which accelerated the catalytic performance of the material (Table 3.1) [13]. Moreover, the enzyme-immobilized hydrophobic organic-inorganic hybrid material showed excellent recycling ability.

The hydrophobic organic-inorganic hybrid material synthesized by a sol-gel reaction and fabricated on an aluminum substrate using a titanium precursor with a hydrophobic silane precursor induced the excellent photocatalytic degradation of organic dyes, such as methylene blue (MB), under UV irradiation [14]. The photocatalytic activity of the silica-doped titania coating on the aluminum substrate was examined by placing the substrate in an aqueous methylene dye solution in the dark for one hour and the photocatalytic degradation was checked by allowing UV light (wavelength range of 200–400 nm) in for various times. The photocatalytic degradation of MB dye was assessed from the adsorption peak intensity of the MB solution before and after UV light irradiation. The silica-doped titania coating on the aluminum substrate showed excellent photocatalytic degradation of MB due to the increased specific surface area, reduced crystallite size, and photocatalytic activity of TiO_2.

Table 3.1 Kinetic resolution of (R,S)-1-phenylethanol catalyzed by immobilized PCL

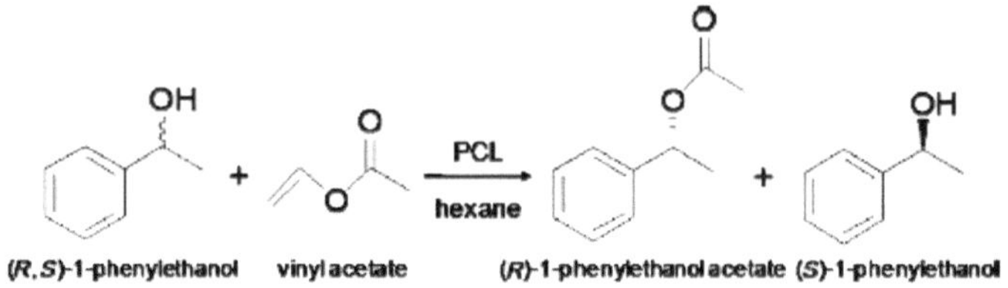

Catalyst	Initial rate [mmol min^{-1} mg$_{protein}^{-1}$]	Conv. [%]	ee[b] [%]
PCL-PMA-0-MS	0.005	8.2	99
PCL-PMA-0.6-MS	0.028	19.5	99
PCL-PMA-1.2-MS	0.033	24.9	99
PCL-PMA-2.4-MS	0.083	41.9	99
PCL-PMA-3.6-MS	0.098	50.0	99
PCL-PMA-4.8-MS	0.088	48.8	99

[a] Typical reaction conditions: (*R,S*)-1-phenylethanol (0.5 mmol) and vinyl acetate (2.0 mmol) were stirred in dry *n*-hexane (5 mL) at 30°C for 6 h. [b] Determined in terms of the ester product.

Source: Reprinted from Ref. [13]. Copyright (2013), with permission from WILEY-VCH Verlag GmbH.

Zhou et al. produced a self-cleaning and hydrophobic organic-inorganic hybrid coating in the presence of photocatalytic TiO_2 nanoparticles and fluorine compounds, which showed excellent hydrophobic stability, was visible-light sensitive, and possessed UV light self-cleaning photocatalytic properties [15]. The photocatalytic and self-cleaning behavior of the polytetrafluoroethylene (PTFE)/fluoroethylene vinyl ether (FEVE) and modified TiO_2-PTFE/FEVE hybrid were assessed using OA as an environmental pollutant.

The hydrophobic TiO_2-PTFE/FEVE hybrid substrate contaminated with OA became hydrophilic under UV light irradiation and the hydrophobicity was recovered on the substrate after several hours in the absence of UV light irradiation (Fig. 3.3). These results highlight the excellent photo-switching and easier self-cleaning properties of the coated substrate. On the other hand, the hybrid substrate in the absence of TiO_2 nanoparticles did not exhibit any photocatalytic activity.

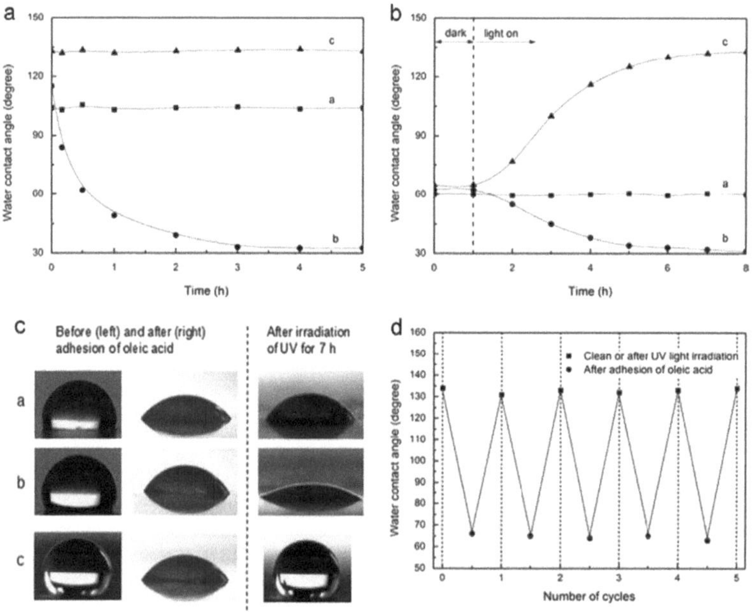

Figure 3.3 (a) Water contact angle (WCA) changes of each test coating sample: (*a*) PTFE/FEVE, (*b*) TiO_2-PTFE/FEVE, and (*c*) modified TiO_2-PTFE/FEVE) under irradiation of UV light. (b) Changes in the WCA by adhesion of oleic acid on the coating surface and following UV light irradiation. (c) Photographic images of water droplet on each test sample before and after adhesion of oleic acid and after irradiation of UV irradiation for 7 h. (d) WCA changes on modified TiO_2-PTFE/FEVE in the five cycles of adhesion of oleic acid and UV light irradiation. Reprinted from Ref. [15]. Copyright (2015), with permission from Elsevier.

The organic-inorganic hybrid quartz crystal microbalance (QCM) surface developed by Wei et al. using zinc oxide (ZnO)-TiO_2 by hydrothermal growth also exhibited excellent self-cleaning, hydrophobic, and photocatalytic activity in the degradation of Rhodamine B (RhB) dye in an aqueous solution [16]. Similarly, the hydrophobic polypyrrole (PPy)/polyvinyl alcohol (PVA)-TiO_2 composite films also showed the photocatalytic degradation of RhB dye in an aqueous solution [17]. More recently, cobalt-iron based TiO_2 ($CoFe_2O_4$-TiO_2) nanocomposite and cobalt-iron-based zeolite Y-silver (zeolite Y-Ag-$CoFe_2O_4$) were synthesized and used for the photocatalytic degradation of azo dye (methyl orange, acid black 1, acid blue 92, and acid brown 14) from an aqueous solution [18, 19]. The synthesized hybrid nanocomposites exhibited excellent photocatalytic degradation of all the azo dyes used in the experiment under UV light irradiation (Fig. 3.4).

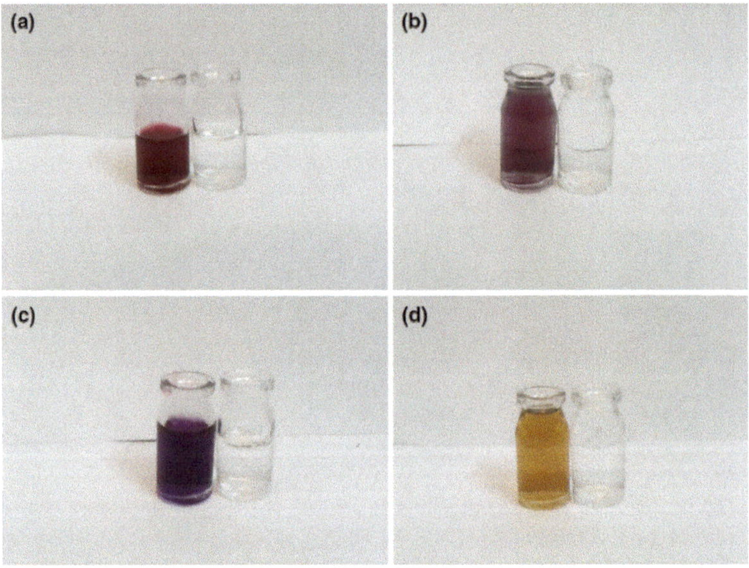

Figure 3.4 Photocatalytic activity of $CoFe_2O_4$-TiO_2 on degradation of (a) methyl orange, (b) acid black 1, (c) acid blue 92, and (d) acid brown 14. Reprinted from Ref. [18]. Copyright (2016), with permission from Springer.

3.2.3 Corrosion Resistance

Corrosion is a major environment problem on many substrates due to the erosion of the substrate under a range of conditions. Corrosion can occur by either chemical or electrochemical reactions and destroy the substrate by oxidation, hydroxylation, and formation of sulfide on the substrate. Corrosion generally takes place on a steel substrate due to electrochemical oxidation and rust formation. Similarly, corrosion also occurs on ceramics or polymer substrates by degradation. The corrosion degradation of steel, ceramic, and polymer substrates can reduce the strength and differentiate the appearance of the substrate.

Recently, hydrophobic and superhydrophobic surfaces were used for anticorrosion applications due to the excellent protection of the hydrophobic materials on the substrate. The hydrophobic coating materials act as a barrier for corrosion protection. Furthermore, the introduction of corrosion inhibitors with hydrophobic materials can reduce rust formation and the degradation of hydrophobic material–coated substrates, as well as increase the lifetime of the coated products.

Ejenstam et al. fabricated a hydrophobic carbon steel surface by spin-coating a hydrophobic solution prepared from a combination of hydrophobically modified silica nanoparticles and polydimethylsiloxane (PDMS) by the catalytic crosslinking of both in the presence of a catalyst and methanol solvent [20]. The fabricated carbon steel surface exhibited stable hydrophobicity due to the hydrophobicity of PDMS and silica nanoparticles. The corrosion resistance of the fabricated substrate was checked against a sodium chloride (NaCl, 3 wt%) solution using a three-electrode system.

The fabricated substrate, saturated Ag/AgCl, and platinum mesh were used as the working PDMS, reference electrode, and counter electrode, respectively. The electrochemical performance, such as electrochemical impedance spectroscopy (EIS) and open-circuit potential (OCP), was checked using an Autolab potentiostat controlled by Nova 1.8 software (Metrohm Autolab B.V.). The OCP and EIS of the fabricated substrates were measured at various exposure times in NaCl solution. The authors checked the corrosion behavior of the fabricated substrate with a bare PDMS-coated substrate as well as various weight percentages of a hydrophobic silica nanoparticle-loaded PDMS surface (Fig. 3.5) [20].

The optical images of Fig. 3.5 clearly revealed the excellent corrosion resistance of the hydrophobic silica nanoparticle–loaded PDMS on the carbon steel surface than the bare PDMS-coated substrate. The results suggest that the excellent corrosion behavior is due mainly to the presence of low-surface-energy hydrophobic silica nanoparticles with hydrophobic PDMS on a carbon steel substrate, which impedes the development of corrosion on the coated substrate. On the other hand, the corrosion resistance was reduced partially by increasing the weight percentage of hydrophobic silica nanoparticles to PDMS on the coated substrate due to the deposition of a large number of loosely connected silica particles on the surface that reduce the corrosion resistance on the coated substrate surface.

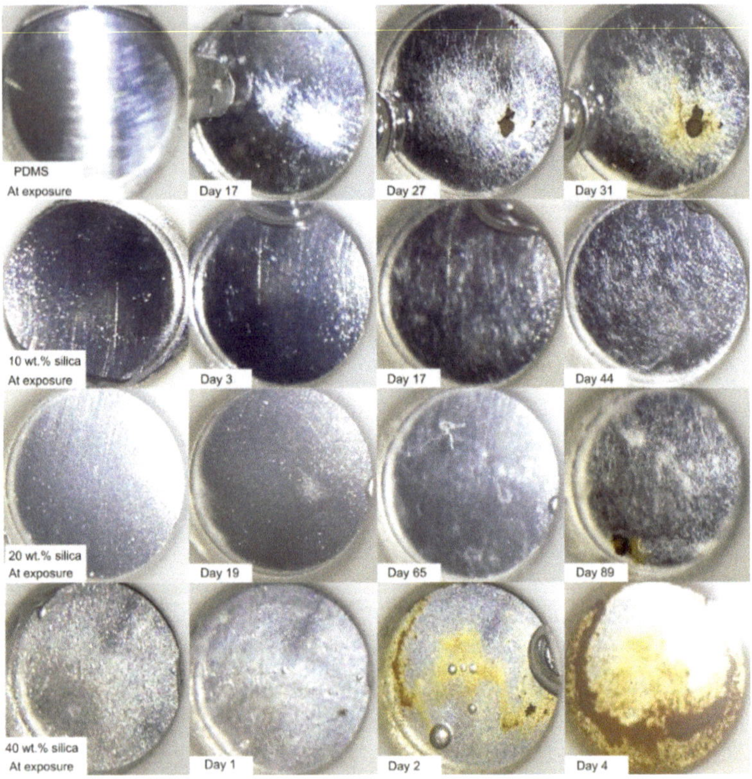

Figure 3.5 Photos of the coated metal surface during exposure to 3 wt% NaCl solution. The diameter of the exposed area is 1 cm, and in most of the photos from the PDMS coating without hydrophobic silica (top row), part of the glass tube used to position the reference electrode can be seen on the left-hand side. Reprinted from Ref. [20]. Copyright (2015), with permission from Elsevier.

Similarly, hydrophobic glass and Q235 steel substrates were fabricated by the spin coating of a perfluorodecyltrichlorosilane

(FDTS)-modified PDMS-ZnO nanocomposite [21]. The hydrophobic steel substrate exhibited excellent anticorrosion activity with a minimum of FDTS in the surface treatment. The fluorine-based low-surface-energy material (LSEM) showed low water adhesion on the surface, which prevented chlorine ion attack and reduced corrosion formation on the coated surface. The ecofriendly hydrophobic patterned graphene surface fabricated on the aluminum alloy substrate also showed good anticorrosion property [22].

The aluminum oxide present on the aluminum alloy substrate can react easily with a Cl^--containing solution and corrode under ambient conditions. On the other hand, the hydrophobic-patterned graphene surface can act a barrier to Cl^- ion attack on the substrate surface and improve the anti-corrosion property of the surface. The hydrophobic aluminum substrate fabricated using a magnesium (Mg)–aluminium (Al) layered double hydroxide (LDH) showed anticorrosion properties. The hydrophobic surface on the Mg alloy substrate was also fabricated as a barrier to corrosion formation and showed excellent anticorrosion properties [23–25].

3.2.4 Scratch Resistance

The term "scratch resistance," or "abrasive resistance," refers to the resistance to scratching or abrasion under severe mechanical stress. The mechanical properties and improved scratch resistance of a substrate surface are important parameters for the preparation of a stable surface on a substrate. The scratch resistance of a substrate depends on several factors, such as adhesion to the material surface, presence of low surface energy in the materials, surface roughness, type of coating method, self-healable property, and the presence of organic-inorganic materials. The scratch resistance or abrasive resistance is improved by the addition of an inorganic nanofiller to organic compounds or polymers. The scratch resistance is also improved by the addition of inorganic nanoparticles, such as SiO_2, TiO_2, and ZrO_2, to the curable resins [26].

The scratch resistance of the coated substrate is generally checked using a Rockwell C diamond scratch indenter. The test is performed by placing the indenter over the coated substrate surface and gently dragging it with a minimum amount of force. The scratch resistance of the coated materials is examined by measuring the depth of surface penetration by the applied stress or load [26]. The scratch resistance of a coated substrate is also tested using the modified scratch resistance test using ball bearings (chromium-alloyed rolling bearing steel) instead of a Rockwell-shaped indenter [27]. Simply, the scratch resistance of a coated substrate is checked from the pencil hardness test value (9B–9H) according to the ASTM standard D3363, where H and B designate the hardness and softness, respectively, properties of the coated material [28]. A substrate with a harder coating with improved scratch resistance shows a higher H value.

Sangermano et al. prepared a UV-curable epoxy coating with hydrophobic and scratch-resistant properties. Initially, the hydrophobically modified silica nanoparticles were prepared by dispersing silica nanoparticles with perfluorooctyltriethoxysilane [26]. The hydrophobic UV-curable coated materials were prepared by mixing the UV-curable resin (cycloaliphatic epoxy resin) with the hydrophobic fluorinated silica nanoparticles. The addition of hydrophobic silica nanoparticles to the UV-curable resin may induce the aggregation of inorganic fillers and enhance the surface roughness and scratch resistance. An increase in the fluorine content in the hydrophobically modified silica nanoparticles may improve the scratch resistance.

Nagappan et al. synthesized an organic-inorganic fluorinated polysiloxane/silica hybrid and deposited it on glass and other substrates, which showed stable hydrophobicity. The prepared substrate also showed good scratch resistance with a pencil hardness of approximately 4H [28]. The hardness was improved slightly by further changing the amount of ethanol in the hybrid dispersion [28, 29]. The hydrophobic glass substrate exhibited excellent antistaining activity. The water and oil-based pen marks can be erased easily from the substrate using soft tissue paper. Similarly, they also synthesized and fabricated hydrophobic organic-inorganic hybrid polyimide-silica nanohybrids, polynorbornene/fluorosilica hybrids, and mercapto-functional polyvinyl chloride metallopolymer glass substrates, and obtained excellent transparency and antistaining properties (Fig. 3.6) [30–32].

Similarly, a hydrophobic surface prepared on a glass substrate by the spin coating of hydrophobic silica nanoparticles obtained from a sol-gel method using methyltrimethoxysilane (MTMS)-

based silica films with 6 wt% PVA exhibited stable hydrophobicity on the substrate [33]. The MTMS-PVA-coated substrate showed a good surface hardness of 4H. Similarly, the highly transparent, high hardness, and excellent scratch-resistant hybrid silica coating was developed by preparing a hydrophobic ormosil sol from the hydrolysis and condensation of diethylenetriamine (DETA)-tetramethyl orthosilicate (TMOS) in an aqueous solvent and a hydrolyzing agent (2-propanol+water) and dip-coated on a polycarbonate substrate [34]. The fabricated substrate exhibited highly stable surface properties with excellent scratch resistance.

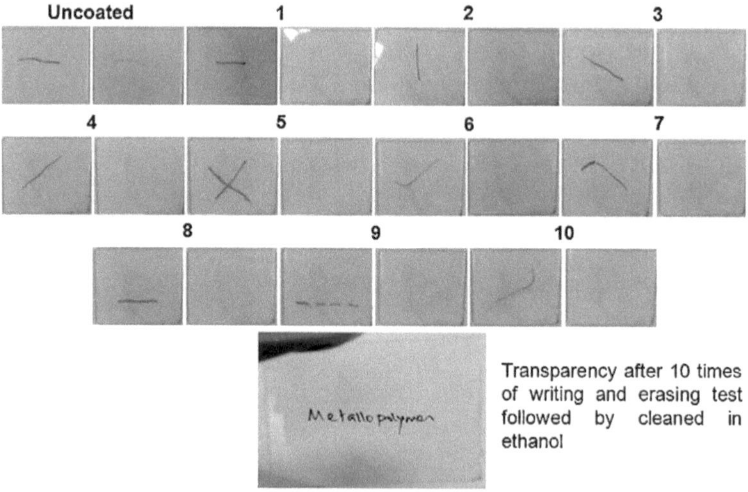

Figure 3.6 Antistaining property of uncoated and functional metallopolymer-coated glass substrate (writing and erasing test using an oil-based pen, 1–10 times). Reprinted from Ref. [32] with permission from The Royal Society of Chemistry.

References

1. Song, B., and Xu, Q. (2016). Highly hydrophobic and superoleophilic nanofibrous mats with controllable pore sizes for efficient oil/water separation, *Langmuir*, **32**, pp. 9960–9966.

2. Qin, H., Gong, T., Cho, Y., Shin, C. Lee, C., and Kim, T. (2015). A simple and economical method using grapheme oxide for the fabrication of water/oil separation papers, *RSC Adv.*, **5**, pp. 57860–57864.

3. Tur, E., Onal-Ulusoy, B., Akdogan, E., and Mutlu, M. (2012). Surface modification of polyethersulfone membrane to improve its hydrophobic characteristics for waste frying oil filtration: radio frequency plasma treatment, *J. Appl. Polym. Sci.*, **123**, pp. 3402–3411.
4. Pei, Y., Han, Q., Tang, L., Zhao, L., and Wu, L. (2016). Fabrication and characterisation of hydrophobic magnetite composite nanoparticles for oil/water separation, *Mater. Technol.*, **31**, pp. 38–41.
5. Mou, F., Pan, D., Chen, C., Gao, Y., Xu, L., and Guan, J. (2015). Magnetically modulated pot-like MnFe2O4 micromotors: nanoparticle assembly fabrication and their capability, for direct oil removal, *Adv. Funct. Mater.*, **25**, pp. 6173–6181.
6. Gu, J., Jiang, W., Wang, F., Chen, M., Mao, J., and Xie, T. (2014). Facile removal of oils from water surfaces through highly hydrophobic and magnetic polymer nanocomposites, *Appl. Surf. Sci.*, **301**, pp. 492–499.
7. Chin, S. F., Romainor, A. N. B., and Pang, S. C. (2014). Fabrication of hydrophobic and magnetic cellulose aerogel with high oil absorption capacity, *Mater. Lett.*, **115**, pp. 241–243.
8. Lin, R., Li, A., Zheng, T., Lu, L., and Cao, Y. (2015). Hydrophobic and flexible cellulose aerogel as an efficient, green and reusable oil sorbent, *RSC Adv.*, **5**, pp. 82027–82033.
9. Jiang, Z. R., Ge, J., Zhou, Y. X., Wang, Z. U., Chen, D. X., Yu, S. H., and Jiang, H. L. (2016). Coating sponge with a hydrophobic porous coordination polymer containing a low-energy CF3-decorated surface for continuous pumping recovery of an oil spill from water, *NPG Asia Mater.*, **8**, p. e253.
10. Qiu, S., Jiang, B., Zheng, X., Zheng, J., Zhu, C., and Wu, M. (2015). Hydrophobic and fire-resistant carbon monolith from melamine sponge: a recyclable sorbent for oil–water separation, *Carbon*, **84**, pp. 551–559.
11. Peng, L., Yuan, S., Yan, G., Yu, P., and Luo, Y. (2014). Hydrophobic sponge for spilled oil absorption, *J. Appl. Polym. Sci.*, **131**, pp. 40886 (1–7).
12. Yamamoto, K., Nohara, Y., and Tatsumi, T. (2001). Synthesis and catalysis of Ti-MCM-41 materials with organic–inorganic hybrid network, *Chem. Lett.*, **7**, pp. 648–649.
13. Liu, J., Peng, J., Shen, S., Jin, Q., Li, C., and Yang, Q. (2013). Enzyme entrapped in polymer-modified nanopores: the effects of macromolecular crowding and surface hydrophobicity, *Chem. Eur. J.*, **19**, pp. 2711–2719.
14. Manjumol, K. A., Mini, L., Peer Mohamed, A., Hareesh, U. S., and Warrier, K. G. K. (2013). A hybrid sol–gel approach for novel photoactive and

hydrophobic titania coatings on aluminium metal surfaces, *RSC Adv.*, 3, pp. 18062–18070.

15. Zhou, Y., Li, M., Zhong, X., Zhu, Z., Deng, P., and Liu, H. (2015). Hydrophobic composite coatings with photocatalytic self-cleaning properties by micro/nanoparticles mixed with fluorocarbon resin, *Ceram. Int.*, **41**, pp. 5341–5347.
16. Wei, Q., Wang, S., Li, W., Yuan, X., and Bai, Y. (2015). Hydrophobic ZnO-TiO_2 nanocomposite with photocatalytic promoting self-cleaning surface, *Int. J. Photoenergy*, **2015**, pp. 925638 (1–6).
17. Cao, S., Zhang, H., Song, Y., Zhang, J., Yang, H., Jiang, L., and Dan, Y. (2015). Investigation of polypyrrole/polyvinyl alcohol–titanium dioxidecomposite films for photo-catalytic applications, *Appl. Surf. Sci.*, **342**, pp. 55–63.
18. Moosavi, S. M., Molla-Abbasi, P., and Haji-Aghajani, Z. (2016). Photocatalyst $CoFe_2O_4$–TiO_2: application in photo-degradation of organic dyes and magnetic nanocomposite preparation, *J. Mater. Sci: Mater. Electron.*, **27**, pp. 4879–4886.
19. Ghanbari, D., Sharifi, S., Naraghi, A., and Nabiyouni, G. (2016). Photodegradation of azo-dyes by applicable magnetic zeolite Y–Silver–$CoFe_2O_4$ nanocomposites, *J. Mater. Sci: Mater. Electron.*, **27**, pp. 5315–5323.
20. Ejenstam, L., Swerin, A., Pan, J., and Claesson, P. M. (2015). Corrosion protection by hydrophobic silica particle-polydimethylsiloxane composite coatings, *Corros. Sci.*, **99**, pp. 89–97.
21. Arukalam, I. O., Oquzie, E. E., and Li, Y. (2016). Fabrication of FDTS-modified PDMS-ZnO nanocomposite hydrophobic coating with anti-fouling capability for corrosion protection of Q235 steel, *J. Colloid Interface Sci.*, **484**, pp. 220–228.
22. Zheng, Z., Liu, Y., Bai, Y., Zhang, J., Han, Z., and Ren, L. (2016). Fabrication of biomimetic hydrophobic patterned graphene surface with ecofriendly anti-corrosion properties for Al alloy, *Colloids Surf. A: Physicochem. Eng. Aspects*, **500**, pp. 64–71.
23. Wang, Y., Zhang, D., and Lu, Z. (2015). Hydrophobic Mg–Al layered double hydroxide film on aluminum: fabrication and microbiologically influenced corrosion resistance properties, *Colloids Surf. A: Physicochem. Eng. Aspects*, **474**, pp. 44–51.
24. Wang, J., Li, D., Liu, Q., Yin, X., Zhang, Y., Jing, X., and Zhang, M. (2010). Fabrication of hydrophobic surface with hierarchical structure on Mg alloy and its corrosion resistance, *Electrochim. Acta*, **55**, pp. 6897–6906.

25. Liu, Y., Lu, G., Liu, J., Han, Z., and Liu, Z. (2013). Fabrication of biomimetic hydrophobic films with corrosion resistance on magnesium alloy by immersion process, *Appl. Surf. Sci.*, **264**, pp. 527–532.
26. Sangermano, M., Perrot, A., Gigot, A., Rivolo, P., Pirri, F., and Messori, M. (2016). Hydrophobic scratch resistant UV-cured epoxy coating, *Macromol. Mater. Eng.*, **301**, pp. 93–98.
27. Sander, T., Tremmel, S., and Wartzack, S. (2011). A modified scratch test for the mechanical characterization of scratch resistance and adhesion of thin hard coatings on soft substrates, *Surf. Coat. Technol.*, **206**, pp. 1873–1878.
28. Nagappan, S., Choi, M. C., Sung, G., Park, S. S., Moorthy, M. S., Chu, S. W., Lee, W. K., and Ha, C. S. (2013). Highly transparent, hydrophobic fluorinated polymethylsiloxane/silica organic-inorganic hybrids for anti-stain coating, *Macromol. Res.*, **21**, pp. 669–680.
29. Nagappan, S., Park, J. J., Park, S. S., Hong, S. H., Jeong, Y. S., Lee, W. K., and Ha, C. S. (2013). Polymethylhydrosiloxane based organic-inorganic hybrids for amphiphobic coatings, *Compos. Interfaces*, **20**, pp. 33–44.
30. Choi, M. C., Sung, G., Nagappan, S., Han, M., and Ha, C. S. (2012). Transparent anti-stain coatings with good thermal and mechanical properties based on polyimide-silica nanohybrids, *J. Nanosci. Nanotechnol.*, **12**, pp. 5788–5793.
31. Sung, G., Choi, M. C., Nagappan, S., Lee, W. K., Han, M., and Ha, C. S. (2013). Polynorbornene/fluorosilica hybrids for hydrophobic and oleophobic coatings, *Polym. Bull.*, **70**, pp. 619–630.
32. Nagappan, S., Park, S. S., Yu, E. J., Cho, J. J., Park, J. J., Lee, W. K., and Ha, C. S. (2013). A highly transparent, amphiphobic, stable and multi-purpose poly(vinyl chloride) metallopolymer for anti-fouling and anti-staining coatings, *J. Mater. Chem. A*, **1**, pp. 12144–12153.
33. Kavale, M. S., Mahadik, S. A., Mahadik, D. B., Parale, V. G., Rao, A. V., Vhatkar, R. S., Wagh, P. B., and Gupta, S. C. (2012). Enrichment in hydrophobicity and scratch resistant properties of silica films on glass by grafted microporosity of the network, *J. Sol-Gel Sci. Technol.*, **64**, pp. 9–16.
34. Yavas, H., Selçuk, C. D. O., Özhan, A. E. S., and Durucan, C. (2014). A parametric study on processing of scratch resistant hybrid sol–gel silica coatings on polycarbonate, *Thin Solid Films*, **556**, pp. 112–119.

Chapter 4

Superhydrophobic Organic-Inorganic Nanohybrids

4.1 Introduction

Superhydrophobic surfaces are of considerable importance in many applications, such as self-cleaning, anti-icing, antibacterial, antireflective, oil sorption and separation, photocatalysts, and solar cells due to the excellent water-repellent properties [1–7]. Superhydrophobic surfaces have been inspired by many biological systems. Superhydrophobic surfaces can show excellent water-repellent behavior, even under severe environmental conditions. This inspiration can lead to the modification or mimicking of the surface properties of various natural superhydrophobic surfaces for use in many applications. Nagappan and Ha briefly reviewed the fabrication of superhydrophobic and magnetic superhydrophobic surfaces using a variety of methods, aspects of fabrication and methods, and their potential applications [8, 9].

Several other authors also reviewed the fabrication of superhydrophobic surfaces and suggested a broad scientific approach to the fabrication of a superhydrophobic surface [10–19]. More recently, self-healing superhydrophobic coatings have been used widely for the fabrication of robust surfaces [20, 21]. The self-healing property of a superhydrophobic surface can reform the

Hydrophobic and Superhydrophobic Organic-Inorganic Nanohybrids
Chang-Sik Ha and Saravanan Nagappan

ISBN 978-981-4774-68-0 (Hardcover), 978-1-351-20607-5 (eBook)
www.panstanford.com

original surface after applying a mechanical stress to the substrate. The solvent-free superhydrophobic coating has also attracted considerable attention in the fabrication of environmentally friendly superhydrophobic coatings [22, 23].

This chapter provides the details of the synthesis and fabrication of superhydrophobic organic-inorganic nanohybrids by various approaches.

4.2 Synthesis of Superhydrophobic Organic-Inorganic Nanohybrids

The synthesis of superhydrophobic materials and superhydrophobic organic-inorganic nanohybrids is quite unusual because the highly hydrophobic materials synthesized from organic-inorganic hybrids can exhibit superhydrophobicity, depending on the substrate selection and pretreatment required to modify the substrate to allow good adhesion of highly hydrophobic materials on the substrate. Some of the materials synthesized using organic-inorganic hybrids exhibit highly hydrophobic properties and can readily form superhydrophobic surfaces on a substrate at room temperature or by a mild thermal treatment after coating the materials on the substrate.

Highly hydrophobic organic-inorganic hybrid materials can be synthesized by various methods, such as click chemistry, emulsion synthesis, surface grafting and modifications, sol-gel synthesis, and various polymerization techniques.

4.2.1 Click Chemistry

Click chemistry is widely used for the synthesis of UV-curable polymers and materials as well as for the synthesis of organic-inorganic hybrid coating materials [24]. The click reaction is carried out generally by reaction of the thiol-based monomers with alkene- and alkyne-based materials under UV-visible light. The thiol-ene click chemistry was initiated by the addition of a photoinitiator, such as azoisobutyronitrile (AIBN), benzophenone, 2,2-dimethoxy-2-phenylacetophenone (DMPA), benzoin ethyl ether, and Irgacure 2100. The thiol-ene hybrid nanofibers were obtained using a range of compositions of acrylo-functional polyhedral oligomeric silsesquioxane (POSS) (P8A), dipentaerythritol pentaacrylate (5A), and pentaerythritol tetrakis(3-mercaptopropionate) (P8A) in the presence of an Irgacure 2100 photoinitiator, polyethylene oxide (PEO) anhydrous ethyl acetate, which showed excellent thermal and mechanical stability [25]. Similarly, UV-curable fluorinated polyhedral oligomeric silsesquioxane (F-POSS) materials have been developed by mixing a mixture of Vinyl-POSS, 1H,1H,2H,2H-perfluorodecanethiol, and DMPA [26]. Chen et al. also developed a flame-retardant cotton fabric by immersing the precleaned cotton fabrics in an aqueous solution of a trilayer of branched poly(ethylenimine) (bPEI) as well as ammonium polyphosphate (APP) solutions [26]. The cotton fabric was modified further by immersing the synthesized FPOSS in an ethanol solution and drying at 60°C. The resulting cotton fabrics showed excellent superhydrophobic, self-healing, and flame-retardant properties. Moreover, the cotton fabrics maintained the above properties for more than 1000 abrasion cycles and showed excellent robustness and multifunctional properties.

Recently, superhydrophobic organic-inorganic hybrid materials were also synthesized by thiol-ene click chemistry and fabricated on various substrates and applied in diverse applications. Chen et al. developed inorganic-organic thiol-ene hybrid materials by mixing hydrophobic silica nanoparticles with 2,4,6,8-tetramethyl-2,4,6,8-tetravinylcyclotetrasiloxane (TMTVSi), pentaerythritol tetra(3-mercaptopropionate) (PETMP), and DMPA in an tetrahydrofuran (THF) solution and spray coating them on a stainless steel mesh surface [27]. The fabricated hybrid substrate was photopolymerized under UV light. The fabricated substrate exhibited superhydrophobic/superoleophilic properties with excellent thermal stability, chemical resistance, and robust mechanical properties. The excellent superhydrophobic/superoleophilic properties of the fabricated substrate can be used to separate various oils from a water surface as well as from oil–water mixtures.

Various thiol and alkene-based organic-inorganic hybrid materials were also synthesized by the click chemistry approach and the superhydrophobic surface on a substrate was prepared by the photopolymerization of thiol-ene compounds with the addition of additional hydrophobic materials. Sparks et al. developed a similar type of superhydrophobic coating material by UV-curable thiol-ene click chemistry and spray-coating the mixture of TMTVSi, triallyl isocyanurate (TTT), DMPA, PETMP, and hydrophobic fumed silica with trimethylsiloxy -Si(OCH_3)$_3$ (Aerosil R972). Surface functionalization was achieved using THF or acetone on glass or other substrates followed by UV curing (5 min) for the photopolymerization of the materials and annealing at 40°C for 2 h [28]. The fabricated substrate showed superhydrophobic (>150°) properties on the glass substrate with a low sliding angle (SA) (2°–5°) and contact angle hysteresis (CAH < 5°).

Similarly, Xiong et al. also developed a superamphiphobic (combination of both superhydrophobicity and superoleophobicity) UV-curable coating by mixing the thiol-ene monomers and other materials [29]. In their study, PETMP was modified with an alkylated fluorine chain (F-PETMP) at the one terminal of the mercapto-functional group using perfluorinated acrylates, such as 1H,1H-perfluoro-n-decyl acrylate or 1H,1H perfluoroheptyl acrylate (Fig. 4.1) [29].

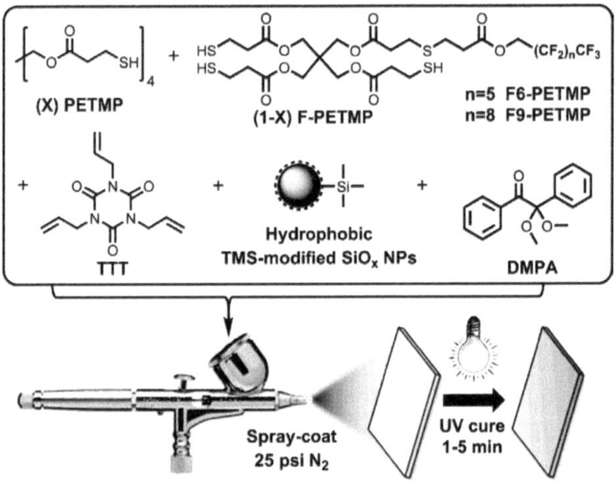

Figure 4.1 Schematic representation of the fabrication of superamphiphobic property by spray deposition and photopolymerization using hybrid inorganic-organic thiol-ene resins laden with hydrophobic silica nanoparticles. Reprinted from Ref. [29]. Copyright (2014), with permission from the American Chemical Society.

The PETMP and F-PETMP were mixed together in the experiment with the alkene chain and with a photoinitiator and hydrophobic silica nanoparticles. The spray-coated solution exhibited excellent superamphiphobicity on a glass substrate after UV photopolymerization. The presence of a fluorine chain at the UV-curable resin is responsible for the superamphiphobic property on the glass substrate. The superamphiphobic surface exhibited excellent water- and various oil-repellent properties on glass and other substrates. These technical advantages of both the combination of water- and oil-repellent properties are quite useful for applying these materials for diverse applications. Moreover, the UV-curable resin by thiol-ene click chemistry is an environmentally friendly, inexpensive, thermally and mechanically stable, and simple process to obtain the robust properties.

4.2.2 Emulsion Synthesis

The emulsion synthesis of hydrophobic coating agents has attracted considerable interest in the fabrication of superhydrophobic organic-inorganic nanohybrid substrates because of the controlled particle size and shape, stability, easy usage, and ability to adhere to various substrates. The emulsion synthesis of superhydrophobic coating materials was developed through simple changes in the formulation of the coating materials. A durable superhydrophobic coating was developed by the emulsion polymerization of a styrene monomer using sodium dodecyl sulfate (SDS) surfactant in the presence of hexadecane (HD) and water [30].

On the other hand, the hydrophobic silica nanoparticles ormosil suspension was synthesized by reacting methyltriethoxysilane (MTES) in methanol in the presence of hydrochloric acid (HCl) followed by the addition of polystyrene (PS) nanospheres (50–200 nm) developed by emulsion polymerization and subsequent stirring of the suspension before applying the coating materials on a substrate. The as-developed organic-inorganic hybrid coating material was spray-coated on a glass substrate and annealed at 350°C for 1 h at a ramp of 1°C/min. The effect of the needle versus crater surface architecture on the coated substrate was analyzed. The emulsion synthesis of the coating material showed hydrophobic to superhydrophobicity on the glass substrate after annealing at 350°C. The emulsion synthesis of PS spheres acts as a sacrificial template within a siloxane matrix. Moreover, the surface morphology of the fabricated substrate depends on the particle sizes of the PS in emulsion synthesis. Increasing the particle size of the PS from 471 to 676 nm also showed the improved surface properties of the coating substrate from hydrophobic to superhydrophobic due to the formation of a hierarchical surface morphology with a crater-like surface roughness. The porosity of the coating material also increased with increasing PS particle size. The increased surface roughness and porosity are responsible for enhancing the repellency of water droplets on the coated substrate. Moreover, the increased particle size of the PS template to 676 nm also showed excellent mechanical stability with a hardness of above 7H (Fig. 4.2). This indicates the robustness of the superhydrophobic coating under mechanical stress, showing that the water repellence is maintained under a strong mechanical force.

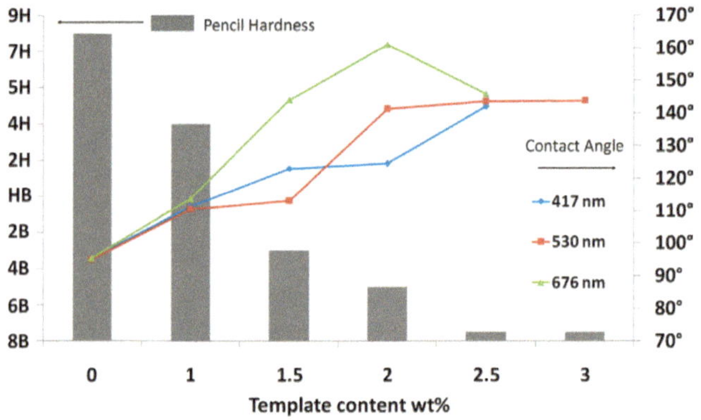

Figure 4.2 The effect of particle size of PS template (417, 530, and 676 nm) under mechanical and surface wettability. Reprinted from Ref. [30]. Copyright (2014), with permission from the American Chemical Society.

On the other hand, ordered macroporous silica particles were synthesized by emulsion synthesis to realize the superhydrophobic surface on a glass substrate (Fig. 4.3) [31]. Cho et al. synthesized PS microspheres by the emulsion polymerization of a styrene monomer and mixed with the hydrophobic silica nanoparticles obtained by the modification of silica nanoparticles in the Stöber method using octadecyltrimethoxysilane (OTMOS). The macroporous particles were synthesized by the emulsification of PS and hydrophobic silica nanoparticles with toluene or HD in water and in the presence of a Pluronic F108 (PF108) surfactant. The surface morphology and porosity of the material on the coated glass substrate was controlled by the coating formations and by calcination of the PS template at 500°C for 3 h in air. The resulting substrate was immersed further in a fluorine-based silane coupling agent (heptadecafluoro-1,1,2,2-tetrahydrodecyl) triethoxysilane (FOS), 1 vol%, using methanol as the dispersing agent. The fluorine-treated glass substrate exhibited superhydrophobicity (166.7°) on the glass substrate with highly porous and uniform pore structures on the silica nanoparticles aggregation.

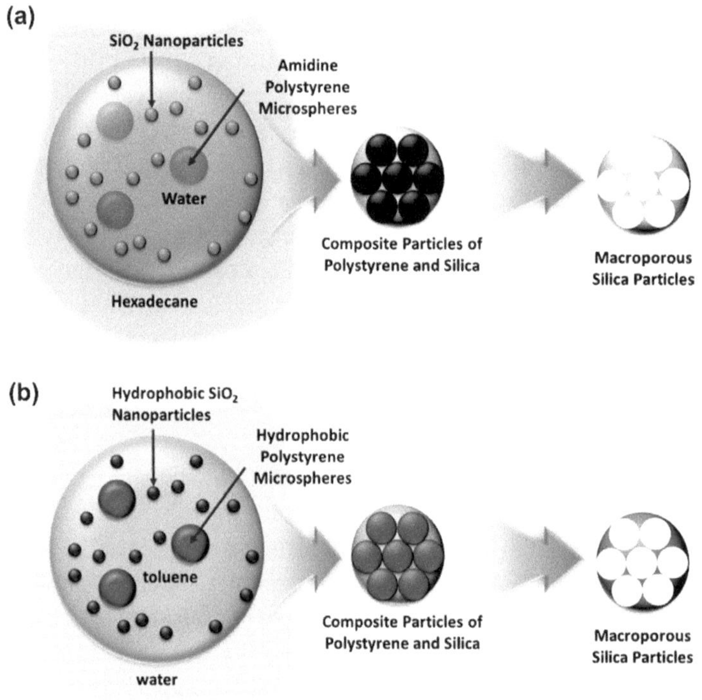

Figure 4.3 Schematic representation of the emulsion synthesis and fabrication of macroporous particles using (a) water-in-hexadecane and (b) toluene-in-water emulsion droplets in the presence of polymer beads and silica nanoparticles. Reprinted from Ref. [31]. Copyright (2012), with permission from Elsevier.

A superhydrophobic organic-inorganic hybrid coating was developed by an emulsion technique using the synthesis of hydrophobic silica nanoparticles mixed with a presynthesized polyacrylate emulsion and spray-coated on a cleaned leather surface, filter paper, and cotton fabrics [32]. The hybrid material-coated leather substrate exhibited superhydrophobicity based on the optimal concentration of hydrophobic silica nanoparticles and polyacrylate emulsion. Moreover the superhydrophobicity can be restored easily even by mechanical stress followed by spray coating on the damaged substrate surface. Emulsion synthesis is simpler and cheaper than other coating formulations used to impart superhydrophobicity on a substrate.

A stable superhydrophobic coating was also fabricated using a fluorinated-polyacrylate (FPA)/silica hybrid coating by free radical-initiated core–shell emulsion polymerization [33]. In this method, emulsion polymerization was carried out using various acrylate monomers, such as methyl methacrylate (MMA), butyl acrylate (BA), hydroxy-propyl acrylate (HPA), and dodecafluoroheptyl methacrylate (DFMA) in the presence of potassium persulfate (KPS) surfactant and emulsifier aqueous solution (DNS-86 dissolved in 40 g H2O). An ethanolic suspension of FPA emulsion/hydrophobic silica nanoparticles was spray-coated on a precleaned glass substrate and dried in an oven at 100°C for 1 h. The surface of the fabricated substrate exhibited stable superhydrophobicity, which may depend on the content of DMFA and hydrophobic silica nanoparticles in the emulsion.

The organic-inorganic hybrid emulsion was also synthesized by reacting aqueous cupric acetate with an ethanolic solution of dodecanoic acid. The emulsion was then spray-coated on a glass or aluminum substrate and dried at room temperature [34]. The fabricated substrate exhibited excellent water repellence due to the formation of a hierarchical surface morphology. In addition, reacting both hydrophilic materials imparted superhydrophobicity (155°) on the substrates with a low SA (<2°) by forming long-chain alkyl groups at the surface of the substrate. In another case, Zhang et al. prepared an organic-inorganic hybrid coating by reaction of a polytetrafluoroethylene (PTFE) emulsion with zinc acetate ($ZnAc_2$) and sodium chloride (NaCl) in the presence of water [35]. A superhydrophobic hybrid substrate was fabricated by dip-coating a glass substrate in the emulsion followed by drying at 100°C for 20 min in an oven and calcination at 370°C for 30 min with subsequent immersion in aqueous acetic acid (1 M) for 30 min and drying at 100°C for 10 min. The fabricated substrate exhibited excellent superhydrophobicity. Moreover, the superhydrophobic substrate was highly stable and robust under severe mechanical stress and abrasion cycles (Fig. 4.4). A similar superhydrophobic surface was also fabricated using various inorganic sources, such as $NiAc_2$ and $CuAc_2$ instead of $ZnAc_2$.

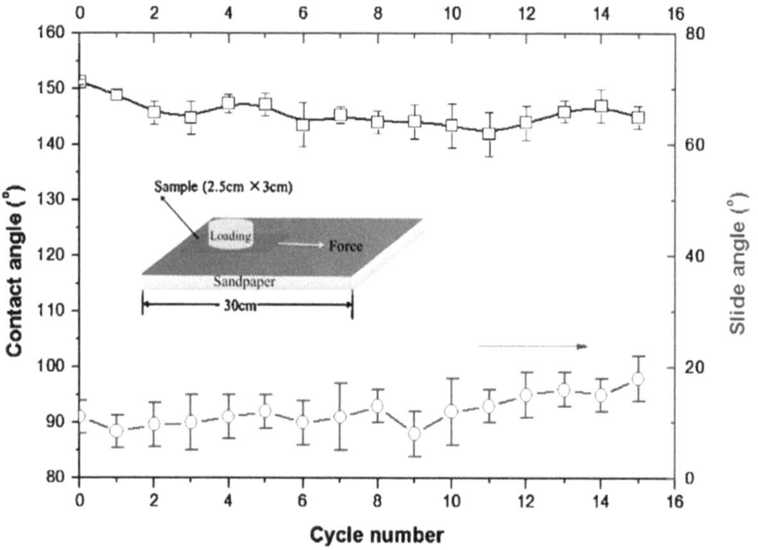

Figure 4.4 The contact angle and sliding angle variations as a function of the number of abrasion cycles. The inset image is the mechanical durability test process setup. Reprinted from Ref. [35]. Copyright (2015), with permission from the Elsevier Ltd.

Recently, Chen et al. developed self-repairable, environmentally friendly, aqueous-based (solvent-free) UV-curable superhydrophobic coating materials [36, 37]. They first synthesized modified silica (SiO_2) and titania (TiO_2) nanoparticles followed by the formation of a shell to the core of the nanoparticles by the emulsion polymerization of styrene (St) and divinylbenzene (DVB) monomers in the presence of (heptadecafluoro-1,1,2,2-tetradecyl) trimethoxysilane (FAS-17a), AIBN, and HD. The surface morphology of the polymer shell was modified further using FAS-17a-modified silica nanoparticles (FMS) (Scheme 4.1) [36].

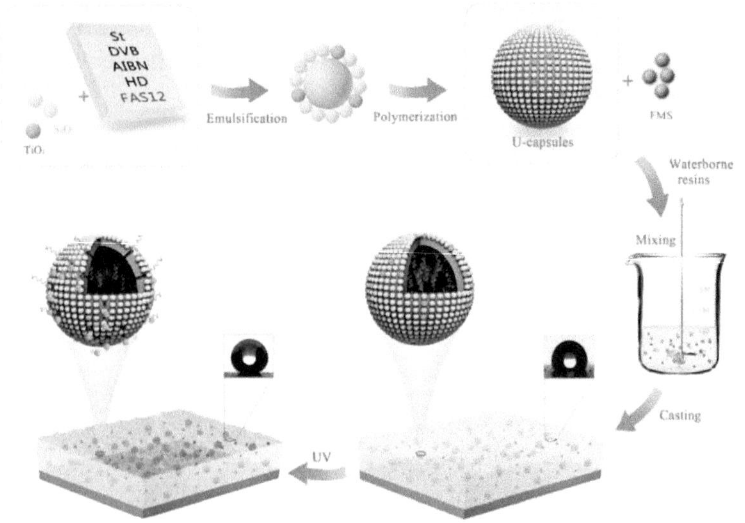

Scheme 4.1 Schematic representation of the fabrication of all-water-based superhydrophobic and self-repairing coatings. Reprinted from Ref. [36]. Copyright (2014), with permission from WILEY-VCH Verlag GmbH.

The effects of the surface morphology were checked in the presence or absence of UV-reactive TiO_2 nanoparticles. The developed coating formulation was applied to a glass substrate by a simple casting technique. The substrate exhibited excellent superhydrophobicity. Furthermore, the surface superhydrophobicity was switched to hydrophilic in the presence of UV light and regained in the absence of UV light. This switchable responsive behavior is due to the photocatalytic activity of TiO_2 nanoparticles and the presence of fluorinated silica nanoparticles that switch the surface property to hydrophilic and hydrophobic based on the responsive nature of the material. The coating formulation is more environmentally friendly due to the aqueous solution as a solvent, and more robust properties of the materials can be useful for diverse applications.

Emulsion synthesis and the fabrication of superhydrophobic organic-inorganic hybrid coatings on glass or other substrate were also attempted using various other materials and for different applications [38–43]. Moreover, continuous interest has also been focused on this topic due to the simple and economically viable approach of the synthesis and fabrication of superhydrophobic coatings. Dong et al. carried out the controlled synthesis of polysilsesquioxane (PSSQ)/PS microspheres by an emulsion technique and studied the superhydrophobicity of the fabricated surface (Scheme 4.2) [43]. First, a highly monodisperse PS latex was synthesized by the dispersion polymerization of styrene in the presence of ethanol, water, polyvinylpyrrolidone (PVP), and AIBN with nitrogen bubbling; the resulting PS particles were modified further by ethyltrimethoxysilane (ETMS) using cetyltrimethyl ammonium bromide (CTAB) and PVP in the presence of water, methanol, and aqueous ammonia. The prepared PSSQ/PS microspheres exhibited superhydrophobicity when fabricated on a glass substrate due to the formation of a hierarchical surface morphology on the substrate. Moreover, the surface roughness and superhydrophobicity depend on the concentration of ETMS in the PS latex. Increasing the concentration of ETMS in the PS latex would increase the surface superhydrophobicity to some level, which is decreased partially by increasing the ETMS concentration further.

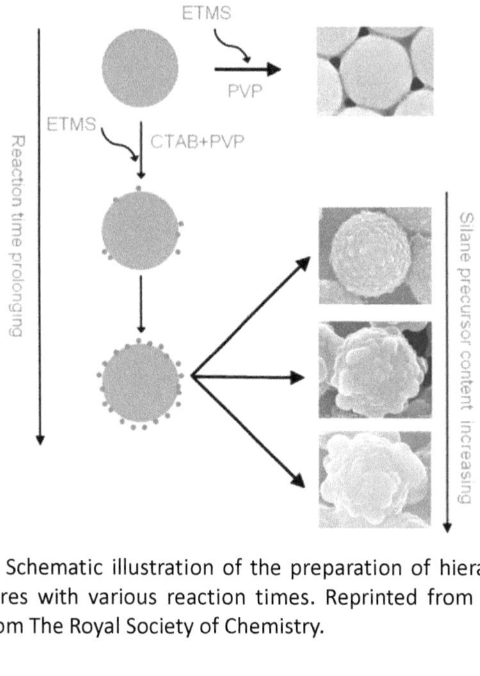

Scheme 4.2 Schematic illustration of the preparation of hierarchical PSSQ/PS microspheres with various reaction times. Reprinted from Ref. [43] with permission from The Royal Society of Chemistry.

4.2.3 Surface Grafting and Modifications

Superhydrophobic organic-inorganic hybrid surfaces are fabricated easily by simple surface treatment of a substrate using a range of hydrophobic grafting agents or crosslinkers. Inspired by the surface properties of various natural sources, simple surface treatment has attracted significant interest to pattern the surface properties of the natural materials.

The superhydrophobic organic-inorganic hybrid copper foam was fabricated by a modification of the copper foam with UV-responsible TiO_2 nanoparticles by dispersion followed by the addition of dopamine (DA) HCl and n-decanethiol (DTH) by adjusting the pH of the solution to 8 using a sodium hydroxide solution and dispersing it for another 9–12 h followed further by washing with deionized water and ethanol and drying naturally at room temperature (Scheme 4.3) [44].

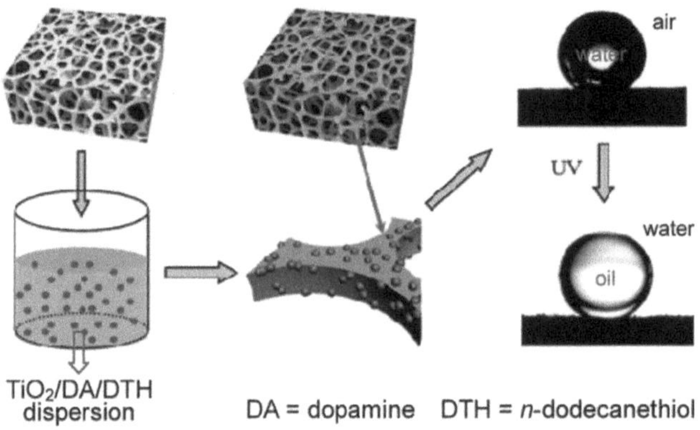

Scheme 4.3 Schematic illustration of the fabrication of UV-switchable superhydrophobic/underwater superoleophobic copper foam. Reprinted from Ref. [44]. Copyright (2014), with permission from the American Chemical Society.

The fabricated copper foam exhibited excellent superhydrophobicity (152.7° ± 2.1°) with a low SA (2.9° ± 1.7°). On the other hand, the superhydrophobicity could be switched to superhydrophilic after a UV (254 nm, 3W) treatment for 2 h due to the photosensitive property of the immobilized TiO_2 nanoparticles.

Furthermore, the modified copper foam also showed underwater superoleophobicity for chloroform.

The surface property of alumina nanoparticles was modified using various branched hydrocarbons as low-surface-energy materials (LSEMs) [45]. The carboxylic acid functional LSEMs (stearic acid, isostearyl acids, and fluorinated carboxylic acids) were reacted at the surface of the alumina nanoparticles, whereas the long chain at the hydrocarbon or alkyl fluorine chains were present at the surface of the modified alumina nanoparticles. The hydrophobically modified alumina nanoparticles were dispersed in 2-propanol (2 wt%) and spray-coated on a range of substrates, such as glass slides, paper, or cardboard, and dried at 40°C–80°C. All the coated substrates were superhydrophobic, but the stability of the superhydrophobic coating on the glass substrate was much weaker than on the other substrate.

A superhydrophobic organic-inorganic hybrid coating was developed on the cotton fabric surface by the γ-radiation-induced graft polymerization of methacryloxypropyl trimethoxysilane (MPTMS) in methanol [46]. The cotton fabric-graft-poly(MPTMS) was modified further by hexamethyldisilazane (HMDS) in acetone and heated under at 70°C for 6 h in a nitrogen atmosphere and washed with ethanol. The electrospun nanofiber membrane developed using polyvinyl alcohol (PVA) showed a uniform morphology with a nonwoven mat and a highly porous structure [47]. The nanofibrous mat was modified further using FAS-17a. The fluorosilane-modified PVA nanofibrous mat showed excellent superhydrophobic and water repellence with a CA and an SA of 158°–153° and 4°–15°, respectively. The fluorinated PVA nanofibrous mat also exhibited excellent chemical resistance against corrosive solutions and polar organic solvents.

The surface grafting and polymerization of the MMA monomer was also carried out on a copper substrate in the presence of the electro-initiated redox activation of aryl diazonium salts [48]. The resulting copper substrate exhibited stable superhydrophobicity without further modification. The uniform and well-ordered silica nanoparticles synthesized by the Stöber method using tetraethoxysilane (TEOS) was modified with different functional

assembled to form a well-ordered nanostructure. Moreover, the fabricated surfaces with various concentrations also exhibited stable superhydrophobicity on the surfaces. The low SA of the fabricated surfaces can repel water droplets from the surfaces more easily and illustrate the self-cleaning behavior of the surfaces.

Yang et al. synthesized a hierarchical flower-like Sb_2WO_6 microsphere by a hydrothermal method in a stainless steel autoclave with a Teflon liner [55]. The self-assembly of Sb_2WO_6 microspheres was achieved by reacting $SbCl_3$ and $Na_2WO_4{\cdot}2H_2O$ in ethanol, distilled water, and sodium hydroxide solutions at 180°C for 24 h. The Sb_2WO_6 nanospheres were obtained by reaction of the materials at the initial times. The self-assembly and growth of the nanospheres by further increasing the reaction time to 24 h may lead to the formation of uniform and complete Sb_2WO_6 microspheres (Fig. 4.5).

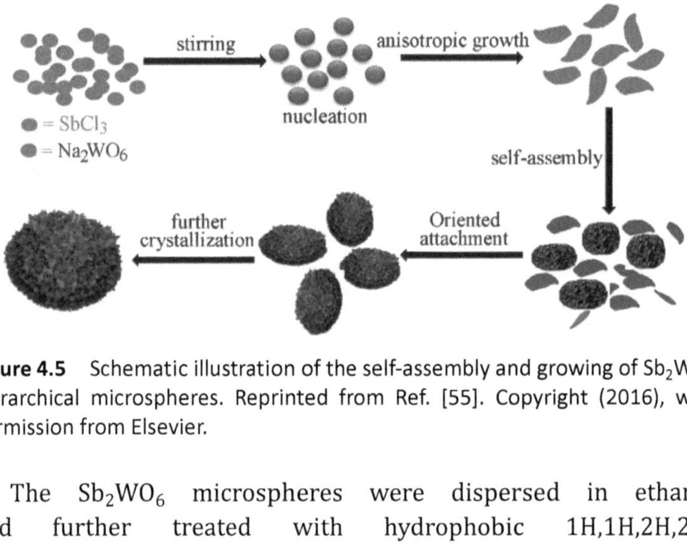

Figure 4.5 Schematic illustration of the self-assembly and growing of Sb_2WO_6 hierarchical microspheres. Reprinted from Ref. [55]. Copyright (2016), with permission from Elsevier.

The Sb_2WO_6 microspheres were dispersed in ethanol and further treated with hydrophobic 1H,1H,2H,2H-perfluorodecyltriethoxysilane (PFOST, 2% v/v) in a methanol solution. Furthermore, the ultrasonically precleaned glass substrate (acetone, ethanol, and deionized water) was dipped into a hydrophobic Sb_2WO_6 microsphere dispersion at room temperature and dried at 80°C in a vacuum for 1 h. The synthesized Sb_2WO_6 microspheres without surface modification on the glass substrate were highly hydrophobic with a CA of 141.2°. On the other hand, the self-assembly of fluorosilane surface-modified hydrophobic

Sb_2WO_6 microspheres on the glass substrate exhibited stable superhydrophobicity (162.68°) due to the formation of a hierarchical surface morphology on a glass substrate with a hydrophobic fluorosilane coating on the surface of the hydrophobic Sb_2WO_6 microspheres. The formation of a dual micronanohierarchical structure of Sb_2WO_6 also exhibited photocatalytic behavior due to the photocatalytic activity of the Sb_2WO_6 microspheres. Moreover, the fabricated Sb_2WO_6 film on the glass substrate also maintained its superhydrophobicity to various acidic and basic solutions.

On the other hand, a superhydrophobic copper surface was fabricated by the self-assembly of multilayer metallic copper films by electrochemical deposition followed by surface treatment with hydrophobic materials [56]. The copper films were grown from copper chloride ($CuCl_2$) and sodium sulfate (Na_2SO_4) solutions and deposited on a copper substrate in a two-electrode system (copper cathode, platinum anode) with a minimum voltage of 0.5 V. The substrate was treated further with an alkaline DA solution followed by 1-dodecanethiol (10 mM in dichloromethane). The resulting copper substrate showed superhydrophobicity (154°).

The superhydrophobic functionalized polyethylene (PE) films were fabricated by the covalent or ionic layer-by-layer self-assembly of functionalized MWCNTs [57]. The MWCNTs were first functionalized covalently with amino-functional polyethyleneimine (PEI) and acylated further with polyacrylic acid (PAA) or a Gantrez copolymer and ooctadecanoic acid. The authors fabricated superhydrophobic surfaces on the PE film in two ways: covalent and ionic layer-by-layer self-assembly. Covalent layer-by-layer self-assembly was carried out by acylation or surface grafting of the amino-functional PEI-modified MWCNTs with the Gantrez copolymer and octadecanoic acid solution On the other hand, ionic layer-by-layer self-assembly was carried out by acylation or surface grafting of the amino-functional PEI-modified MWCNTs with PAA and octadecanoic acid solution. The surface wettability of the PE films obtained by covalent and ionic layer-by-layer self-assembly of the hybrid materials showed CAs of 165° and 155°, respectively. The PE films fabricated by covalent layer-by-layer self-assembly were more robust to various pH conditions and maintained their superhydrophobicity. On the other hand, the PE films fabricated by ionic layer-by-layer self-assembly showed a labile nature to acidic pH conditions.

Hierarchical ZnO nanowires were synthesized by the hydrothermal self-assembly of zinc acetate dehydrate ($Zn(CH_3COO)_2 \cdot 2H_2O$) in a neutral aqueous solution on a Si(111) substrate at 100°C for 12 h [58]. The surface properties of the ZnO nanowires were modified using heptadecafluorodecyl trimethoxysilane (HTMS). The surface-modified ZnO nanowires on the Si(111) substrate exhibited superhydrophobicity due to the formation of a micronanohierarchical reticulate surface morphology. The formation of this surface morphology was attributed to the aggregation of ZnO nanowires and the rapid evaporation and contraction of water drops. The fluorosilane-treated reticulate ZnO nanowires showed stable superhydrophobicity (170° ± 1°) and a low SA (2°).

Recently, Du et al. briefly reviewed 3D self-supporting porous magnetic assemblies (3D-SPMAs) due to the combination of porous assemblies and magnetic materials [59]. The 3D-SPMAs are useful in many applications because of their highly porous structure and magnetic-responsive behavior. The highly porous structure of the 3D-SPMAs is quite useful for the fabrication of magnetic-responsive superhydrophobic sponges or foams, which can be used in several applications. Nagappan and Ha reviewed the emerging trends in superhydrophobic-based magnetic surfaces [9]. The self-assembly of magnetic and other molecules in a porous material may lead to the easily formation of superhydrophobic surfaces and exhibit magnetically responsive behavior. Yu et al. prepared a superhydrophobic magnetic titanium dioxide foam by first modifying the ordinary PE shock absorption foam with a self-assembly of oleic acid and magnetic nanoparticles via a solvothermal approach followed by further modification of the foam by immersion in a titanium dioxide solution and methyltrimethoxysilane (6% v/v) solution [60]. The prepared foam exhibited stable superhydrophobicity with a water contact angle (WCA) of 152.1° ± 1.2°.

4.3.1.2 Sol-gel method

The sol-gel method is used widely for the fabrication of superhydrophobic coatings due to the simple and easy surface modification of various metal precursors by a hydrolysis and condensation reaction at room temperature. The metal precursors can be hydrolyzed easily under acidic or basic or a combination of acidic and basic conditions using the sol-gel method. Organic silicates and some metal precursors with the general formula, $R'_nM(OR)_{4-n}$ or $(RO)_3MR'(OR)_3$, are used generally as the metal organic precursors for the preparation of organic-inorganic hybrid materials. The hydrolyzed and condensed metal precursors of the organic-inorganic hybrid materials by the sol-gel method can be used in many applications owing to their excellent chemical and thermal stability, and corrosion resistance on various substrates. The gelation of metal precursors in the sol-gel method depends on several factors, such as pH, water–precursor ratio, and types of metal precursor, reaction times, temperature, and solvent.

A sol-gel method was used to fabricate a multifunctional coating with antireflection and self-cleaning properties on a soda-lime glass substrate. The sol-gel method was carried out by mixing TEOS or titanium butoxide (n-BuTi) in the presence of Pluronic F127 (PF127) surfactant in the presence of ethanol, water, and HCl solution [61]. The organic-inorganic hybrid solution formed a stable superhydrophobic surface on a glass substrate with low hysteresis and SA.

Zhang et al. prepared superhydrophobic organically modified silicate (ormosil) thin films using a template-free sol-gel method. The prepared thin films exhibited excellent antireflection properties with a refractive index value varying from 1.23 to 1.13 [62]. The superhydrophobic ormosil thin films were prepared in three ways: (i) hydrolysis and (ii) gel formation of silane precursors using ethanol and water in the presence of acid, base, or a combination of both acid-base catalysts, followed by (iii) hydrophobic surface treatment using HMDS of the silica nanoparticles or no such treatment (Fig. 4.6). The hydrophobic treatment of the silica nanoparticles involved dip coating on a glass (BK7) substrate followed by drying at 160°C for 2 h. The resulting thin film showed a uniform thickness and excellent transparency (99.6% to 99.8%) as well as superhydrophobicity on a glass substrate.

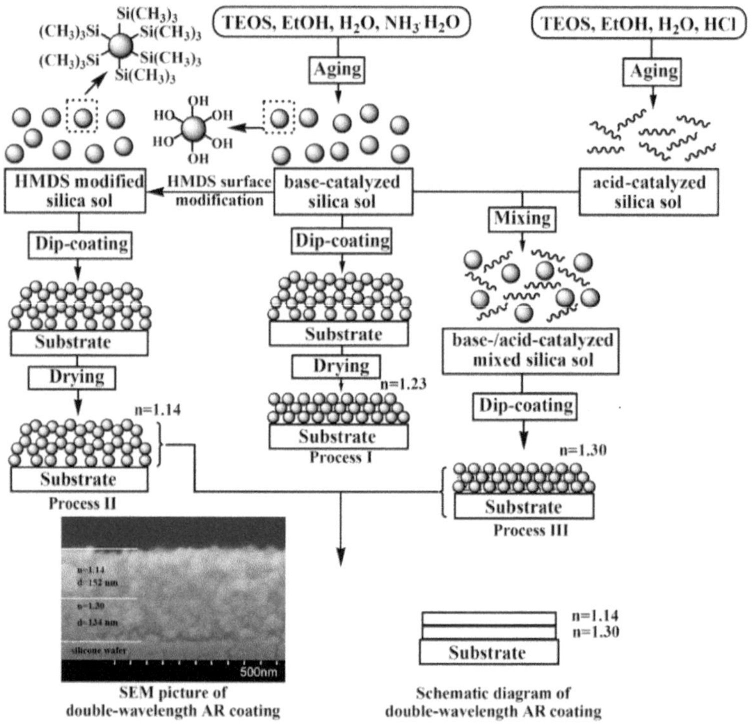

Figure 4.6 Schematic representation of sol-gel coating made by using various protocols. Reprinted from Ref. [62]. Copyright (2013), with permission from WILEY-VCH Verlag GmbH.

The gelation by hydrolysis and condensation in the sol-gel method may depend on several factors, such as the choice of silane precursor, temperature, pH, reaction medium, solvent, and water–silane ratio. A template-free mesoporous superhydrophobic powder was prepared by a sol-gel method using polymethylhydrosiloxane (PMHS) and TEOS with a base catalyst (NaOH) in the presence of ethanol and water by hydroxylation and condensation [63]. In a similar manner, Nagappan et al. obtained a superhydrophobic powder in one-pot by a sol-gel method without the use of TEOS [64]. The powder showed very high surface roughness, superhydrophobicity (174.5° ± 1.0°), hierarchical surface morphology, and mesoporous structure (Fig. 4.7). The superhydrophobic powder became superhydrophilic and reduced the porous structure by calcination at 600°C. The superhydrophobic and superhydrophilic powders exhibited pH-sensitive methylene blue (MB) adsorption behavior.

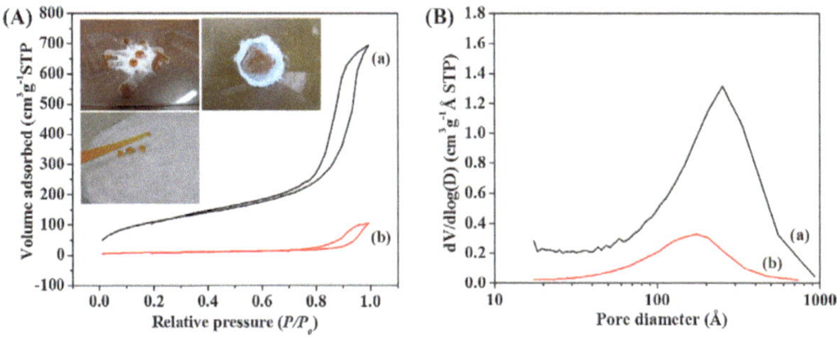

Figure 4.7 (A and B) BET surface areas and pore size distributions of (a) PMHS and (b) PMHS 600°C (the inset images show the superhydrophobic and superhydrophilic properties of PMHS and PMHS 600°C. Reprinted from Ref. [64]. Copyright (2015), with permission from Elsevier.

The superhydrophobic polymethylsiloxane was also synthesized using different concentrations of sodium hydroxide and modified with functional silica ormosils [65, 66]. The thermally stable superhydrophobic hybrid powder and substrate were also obtained by the in-situ hydroxylation and condensation of polymethylsiloxane with graphene oxide (GO) and MWCNTs [67, 68]. The template-free superhydrophobic surface in the sol-gel process for an antireflective coating was also prepared using TEOS and HMDS precursors in the presence of an acid or base catalyst [62]. The silica thin-film surface showed good transparency of 99.6% and 99.8% at 532 nm and 1064 nm, respectively, and an ultralow refractive index (1.23 to 1.13).

A superhydrophobic foam type material was prepared by a sol-gel method and the switchability of the superhydrophobic bulk materials was examined by thermal annealing [69]. The superhydrophobic bulk samples were prepared by the hydrolysis and condensation of MTES and TEOS in the presence of dimethylformamide (DMF) and acid catalyst. The solution was aged for approximately 7 days, and the bulk gel was then washed with methanol via a solvent exchange method and heat treated to obtain the bulk superhydrophobic foam. The superhydrophobic foam was also prepared using phenyltriethoxysilane (PTES) instead of MTES in the same procedure.

The bulk sample showed stable superhydrophobicity (CA, over 155°) up to a certain temperature (400°C) at which the surface property changed from superhydrophobic to superhydrophilic.

The MTES-treated superhydrophobic bulk foam maintained its superhydrophobicity up to 390°C; the surface became superhydrophilic when the annealing temperature was increased to 400°C. This is due to the partial changes in the surface functionality and the degradation of hydrophobic materials on the surface of the silicone backbone. The addition of PTES to TEOS can maintain the stability of the superhydrophobic behavior of the bulk foam up to 550°C due to the enhanced thermal stability of the phenyl ring at the bulk surface [64–69]. Moreover, the enhancement may depend on the concentration of PTES with the TEOS mixture. The superhydrophobic coating prepared by the sol-gel method using MTES, 3-glycidyloxypropyl trimethoxysilane (GLYMO), and FAS-17b in the presence of ethanol, deionized water, and itaconic acid showed stable superhydrophobicity and anti-icing properties [70].

A transparent superhydrophobic and self-cleaning coating on a glass substrate was prepared by the hydrolysis and condensation of (heptadecafluoro-1,1,2,2-tetrahydrodecyl) trimethoxysilane (FAS-17c) in the presence of an ethanol, deionized water, and ammonia solution. First the deionized water (1 mL), ethanol (25 mL), and ammonia (0.5 mL) solutions were stirred for 5 min followed by the slow addition of FAS-17c and stirring for another 5 min and the precleaned glass substrate was immersed (10, 100, 300, and 600 s) into the hydrophobic solution and dried overnight at room temperature [71]. The prepared sample exhibited excellent superhydrophobicity on the glass substrate. The surface roughness of the coated glass substrate was increased gradually from 100 nm, 130 nm, and 210 nm by increasing the immersion time and decreased (170 nm) partially by further increasing the immersion time. The surface CA also increased gradually from hydrophobic to superhydrophobic (CA, 127°, 141°, 169°) and the SA (SA, 31°, 19°, <5°) decreased with increasing immersion time (10, 100, and 300 s). The formation of superhydrophobicity on the glass substrate was attributed to the formation of good surface roughness and a wrinkled hill-like morphology and highly branched hydrophobic fluorosilane materials at the surface. Moreover, the prepared superhydrophobic glass substrate exhibited good transparency (82%) (Fig. 4.8).

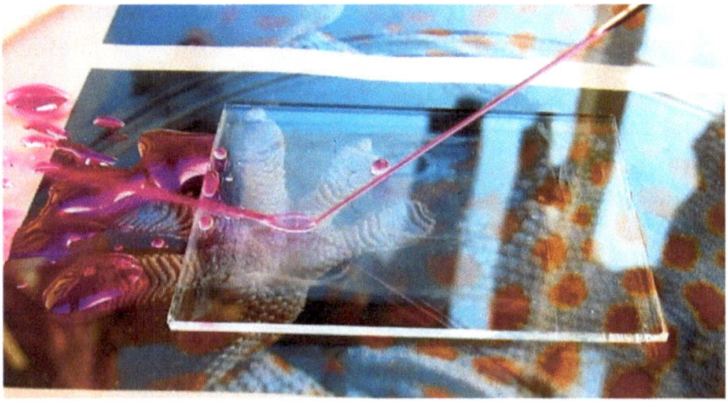

Figure 4.8 Superhydrophobic coating on a glass substrate made by sol-gel method. Reprinted from Ref. [71]. Copyright (2015), with permission from Elsevier.

Similarly, transparent superhydrophobic coatings by the sol-gel method were prepared using different silane precursors and surface modifier for the hydrolysis and condensation of the materials [72, 73]. A superhydrophobic membrane and textile fabrics were also prepared by a sol-gel method using different silane precursors. These prepared samples maintained their superhydrophobicity on the substrate and their surface properties were altered on the basis of the chemical nature of the materials used in the sol-gel method [74–76].

Das et al. fabricated zirconia-based superhydrophobic organic-inorganic hybrid coatings on cotton fabrics by a sol-gel method [76]. First, they prepared an acetyl functional zirconia sol by stirring zirconium(IV) n-propoxide (ZP) in an n-propanol (70%), acetylacetone, 1-butanol, and nitric acid solution followed by the surface modification of functional zirconia using 0.2 g of FAS-17b dissolved in 20 g of 1-butanol. The cotton fabric was immersed in a fluorosilylated zirconia solution for 2 h and then dried at 60°C for 30 min and then at 120°C for 1 h. The fabricated cotton fabrics exhibited excellent superhydrophobicity (163° ± 1°), as well as low CAH ≈ 3.5°, self-cleaning, and antistaining behavior.

4.3.1.3 Solution immersion

Solution immersion is the most prominent method to fabricate a superhydrophobic coating due to easy and simple method of surface modification on various substrates. The superhydrophobic coating can be obtained using a one- or two-step immersion process. In most cases, the substrates are pre-etched in a strong basic or acidic solution to form a rough surface morphology on the substrate coated with thin films of hydrophobic materials to produce a superhydrophobic surface on the substrate. Various hydrophobic organic or inorganic-organic materials have been used to develop superhydrophobic surfaces on various substrates by a simple immersion technique. Moreover, the immersion technique is environmentally friendly and reduces the time of preparation and materials consumption, and energy.

An environmentally friendly, facile, and low-cost hydrophobic melamine-formaldehyde (MF) sponge was fabricated using a solution immersion technique [77]. The highly hydrophobic sponge was prepared by immersing the MF sponge in lignin and carbodiimide-modified diphenylmethane diisocyanate (MMDI) dissolved in a THF solution for 2 s, followed by drying at ambient temperature at 2 h. The shell thickness of the hydrophobic lignin on the MF sponge surface was in the range of 0.1 to 0.4 cm. The lignin-modified MF sponge exhibited hydrophilicity (68°) at low lignin concentrations (0.005 g/mL) due to the low modification level of lignin on the MF sponge surface. Increasing the lignin concentration to a certain level on the MF sponge surface resulted in an increase in hydrophobicity (145.9°); further increases in the lignin concentration resulted in a decrease in surface hydrophobicity (129°) on the MF sponge surface. Similar trends regarding the surface property were observed by adding MMDI with different MMDI to lignin ratios. The surface wettability of lignin-MMDI-modified MF sponge can show stable surface hydrophobicity at an MMDI-to-lignin ratio of 1:1 (145.9°) and 2:1 (138.5°). The fabricated hydrophobically modified MF sponge can be used for the sorption and separation of oils from water.

Liu et al. fabricated an efficient superhydrophobic and superhydrophilic stainless steel mesh by the one-step surface modification of the stainless steel by immersing the precleaned stainless steel substrate in a 1.5 M $CuCl_2$ and 0.5 M HCl solution for different immersion times (2, 5, 7, 10, 15, 20, and 30 s) at room temperature and dried at 70°C for 30 min [78]. The stainless steel surface was immersed further in a 0.1 M ethanol solution of stearic acid for 30 min and rinsed in an alcohol solution and dried at 60°C for 15 min. The surface morphology of the stainless steel mesh immersed for various times in a $CuCl_2$ solution showed the formation of micronanohierarchical structures on the surface. The smooth surface of the stainless steel mesh changed to a highly hierarchical rough surface by increasing the immersion time in a $CuCl_2$ solution (Fig. 4.9).

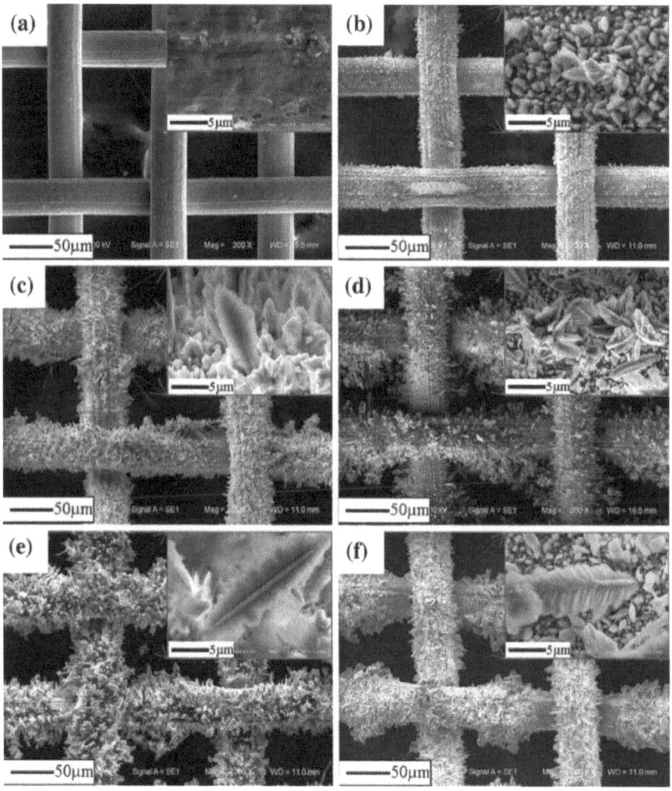

Figure 4.9 Surface morphology of the stainless steel mesh at different reaction time in an aqueous solution of 1.5 M $CuCl_2$ and 0.5 M HCl. SEM images of newly prepared stainless steel samples after immersion times of (a) 0 s, (b) 2 s, (c) 5 s, (d) 7 s, (e) 15 s, and (f) 30 s. The insets show magnified images of corresponding surfaces. Reprinted from Ref. [78]. Copyright (2016), with permission from Elsevier.

This is because of the presence of a high concentration HCl solution during the immersion as well as the uniform deposition of copper crystals on the substrate. Moreover, the crystallinity of copper changes from a needle to a nanowire morphology, which induces surface roughness by increasing the immersion time. The surface was modified further by immersion in a stearic acid solution. A thin hydrophobic coating was deposited on the rough surface, and a superhydrophobic (153° ± 3°) surface was obtained within 15 s [78]. Zhu et al. also fabricated a superhydrophobic surface in a stainless steel mesh by a one-step solution immersion approach using stearic acid as the surface modifier [79]. The stearic acid-modified stainless steel substrate showed hydrophobic to superhydrophobicity according to the concentration of the stearic acid solution. The surface superhydrophobic behavior increased from 143° ± 1° to 160° ± 1.0° with increasing stearic acid concentration.

A superhydrophobic steel plate (Q235) was also fabricated using a simple low-cost method. First, the steel plates were polished mechanically using silicon carbide (SiC) sandpaper and then washed in ethanol and dried at 60 C for 5 min [80]. The steel plates were immersed in a solution containing a mixture of copper(II) sulfate pentahydrate ($CuSO_4 \cdot 5H_2O$, 2.5 g/L), selenous acid (H_2SeO_3, 5 g/L), nickel nitrate ($NiNO_3$, 0.5 g/L), $Zn(HPO_3)_2$ (1 g/L), citric acid ($C_6H_8O_7$, 2.3 g/L), phosphorous acid (H_3PO_3, 2.2 mL/L), and sodium lauryl sulfate ($C_{12}H_{25}SO_4Na$, 0.1 g/L) for 8–15 min and at 60°C for 5 min. The dried steel plate was immersed further in a stearic acid (5 g/L) solution for 10 min at 60°C and dried at 60°C for 10 min. The fabricated steel plates showed a hierarchical surface morphology to provide superhydrophobicity (151°) with low SA (6°) at the initial time of immersion (approximately 10 min). After increasing the immersion time to 16 min, the superhydrophobic CA was increased to 158° and the SA was 1°. The steel substrate exhibited stable superhydrophobicity at various pH values from 1 to 14.

A novel environmentally friendly superhydrophobic surface was prepared from recycled paper [81]. Initially, stearic acid–modified magnesium hydroxide ($MgOH_2$) and DA–silica trimethylsilyl–modified gel powder were prepared separately and used for the preparation of a flame-retardant and superhydrophobic surface in recycled waste paper [81]. The waste paper fragments were immersed in deionized water and stirred vigorously at 80°C for 3 h, and the cellulose pulp was filtered and redispersed in an ethanol solution to modify the surface properties of the cellulose pulp by adding different compositions of stearic acid–modified magnesium hydroxide ($MgOH_2$) (STA-MH) and DA–silica trimethylsilyl–modified gel powder. The suspension was dispersed for a further 30 min; the modified pulp was filtered, dried at 60°C for 2 h in air, and pressed into paper under 5 MPa using a tablet press. The resulting flat and flexible recycled paper exhibited superhydrophobicity and excellent flame-retardant properties (Fig. 4.10). The superhydrophobic recycled paper also showed excellent self-cleaning property and chemical durability against strong acid (1 M HNO_3) and base (1 M NaOH).

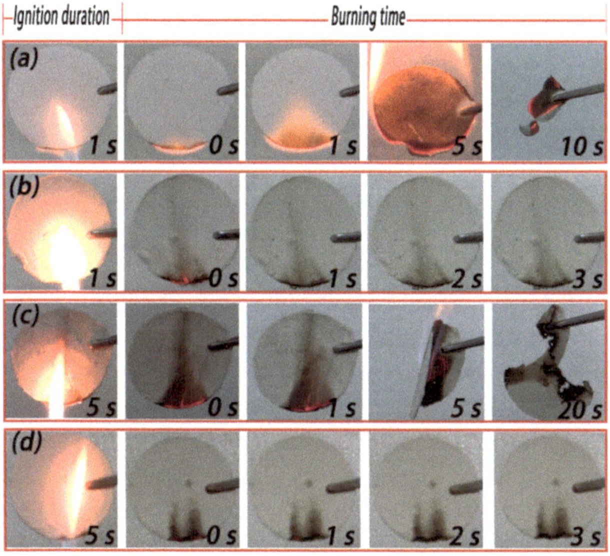

Figure 4.10 (a) The combustion process of cellulose fiber paper with 1 s ignition duration. (b) The combustion process of superhydrophobic and flame-retardant recycled paper with 50% STA-MH with 1 s ignition duration. (c) The combustion process of superhydrophobic and flame-retardant recycled paper with 50% STA-MH with 5 s ignition duration. (d) The combustion process of superhydrophobic and flame-retardant recycled paper with 83.3% STA-MH with 5 s ignition duration. Reprinted from Ref. [81]. Copyright (2016), with permission from Elsevier.

Micronanostructured TiO_2 particles were synthesized using butyl titanate in the presence of ethanol, deionized water, and sodium hydroxide solutions [82]. The micronanostructures were produced by preparing two solutions, such as butyl titanate in an ethanol solution as well as an ethanol-deionized water solution with the pH of the solution adjusted to pH 9 using 1 mol/L NaOH into deionized water. Both mixtures were added together to obtain micronanostructured TiO_2 particles. The surface properties of the micronanostructured TiO_2 particles were modified with an amino-functional coupling agent (aminopropyltriethoxysilane). A superhydrophobic filter paper was prepared by immersing the filter paper in aminosilane-modified TiO_2 particles followed by immersion in an epoxy resin solution and octadecyltrichlorosilane (OTS) and drying at room temperature. The filter paper surface before and after surface modification was superhydrophilic (5° ± 0.5°) and superhydrophobic (153° ± 1°), respectively. The prepared superhydrophobic filter paper exhibited good chemical stability and UV resistance.

Chen et al. fabricated organic-superhydrophobic cotton fabrics using an immersion technique from ammonium polyphosphate (APP)/branched poly(ethylenimine) (bPEI) and fluorinated polyhedral oligomeric silsesquioxane (F-POSS) [83]. The cotton fabrics were immersed in an aqueous bPEI solution (4 mg/mL) for 20 min followed by the removal of an excess of water and further immersion in an aqueous solution of APP dispersion (20 mg/mL) for 1 h and washed with water and the fabrics were overnight dried at 60°C. The fabricated cotton fabrics showed intumescent flame-retardant properties after being modified further with F-POSS (10 mg/mL) by immersion for 5 min and dried at 60°C. The resulting cotton fabrics were superhydrophobic and exhibited self-healing behavior to a superhydrophobic coating by maintaining the intumescent flame-retardant property. Moreover, the self-healing superhydrophobic behavior and intumescent flame-retardant properties were maintained in the fabricated cotton fabrics for more than 1000 cycles of abrasion under a pressure of 44.8 kPa.

Robust and environmentally friendly superhydrophobic fabrics were also fabricated by a simple solution immersion method using fluorine-free coating materials [84]. A superhydrophobic fabric coating was obtained by a two-step solution immersion approach using a silica ormosil coating followed by a PDMS coating. The silica ormosil was synthesized by hydrolyzing hydrophobic methyltrimethoxysilane (MTMS) in methanol under acid (oxalic acid) and base (ammonium hydroxide, 28%) catalysts. The solution was aged for gelation for 48 h and fractioned at various concentrations in methanol and used for the surface modification of the fabric. The fabric was immersed in the hydrophobic ormosil suspension for 30 s and dried at room temperature and modified further by immersion in a PDMS/curing agent (Sylgard 184B) (10:1) solution prepared in tetramethylene oxide solution and cured at 80°C for 2 h. The fabricated fabrics exhibited hydrophobicity to superhydrophobicity, depending on the concentration of the fraction of ormosil in methanol. The obtained superhydrophobic fabrics showed excellent separation behavior for various oils from a water surface.

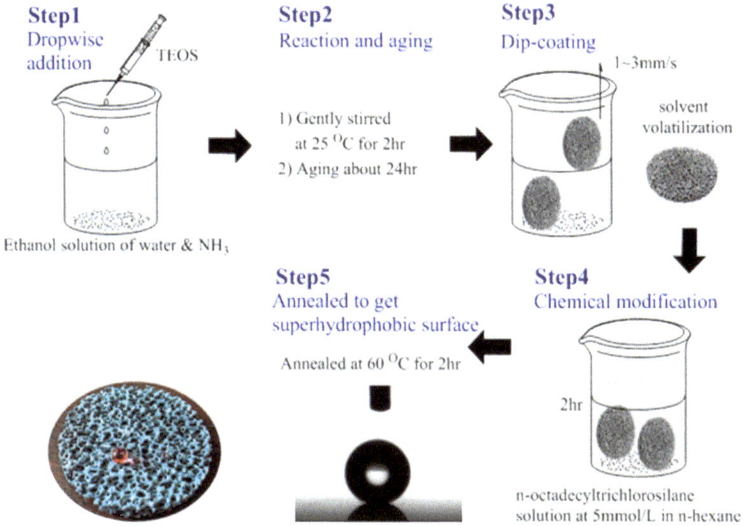

Figure 4.11 Schematic diagram of experimental procedures to prepare superhydrophobic SiC foam. Reprinted from Ref. [85]. Copyright (2016), with permission from the American Chemical Society.

A superhydrophobic ceramic foam material was fabricated in several steps by modifying the surface properties of foam silicon carbide (SiC) materials using various immersion techniques [85]. The superhydrophobic foam was fabricated by first preparing a silica sol using TEOS (3 mL), various volumes of ammonium hydroxide (3, 2.5, or 2 mL), and ethanol (50 mL) and stirred at 25°C for 2 h, followed by aging for 24 h. The resulting sol was dispersed and the

SiC foam was immersed in the sol for 30 s and dried naturally. The immersion cycle was repeated for three to five cycles and further gelation was continued by drying the foam at 60°C–100°C for 2 h. The modified SiC foam was hydrophobized by immersing in an n-octadecyltrichlorosilane/n-hexane (5 mmol/L) solution for 2 h at 25°C and dried at 60°C for 2 h (Fig. 4.11). The resulting modified SiC foam was hydrophobic (140° and 125°) and superhydrophobic (155°). The surface of the foam was modified by changing the particle size and immersion type of the silica sol as well as by hydrophobic surface treatment.

4.3.1.4 Electrochemical deposition

Superhydrophobic surfaces fabricated by electrochemical deposition are useful in anticorrosion applications due to the uniform deposition of organic and inorganic hybrid materials thin coatings on a range of conducting substrates. The coated substrates can exhibit hydrophilic or hydrophobic to superhydrophobic surface properties according to the selection of materials used for electrochemical deposition [86]. Wang et al. fabricated superhydrophobic coatings on copper (Cu, T2), aluminum (Al, 1100), and steel (Q235B) substrates by an electrodeposition method [87]. The superhydrophobic substrates were fabricated by a simple electrochemical deposition process of zinc ions and DA in an electrolyte solution. The surface wettability of the coated substrate depends on the electrochemical deposition time and voltage. The copper (Cu, T2), aluminum (Al, 1100), and steel (Q235B) substrates were abraded with commercial sandpaper and precleaned before applying the coating materials.

The coating solution was prepared by mixing zinc sulfate heptahydrate ($ZnSO_4$ $7H_2O$, 0.02 mol/L), potassium chloride (KCl, 0.1 mol/L), and DA (0.0065 mol/L) in distilled water, and the pH of the solution was adjusted to 6. The coating solution was used as an electrolyte, whereas copper, aluminum, and steel substrates were used as an anode in a two-electrode electrochemical cell. The deposition height and temperature were constant at 2 cm and 40°C, respectively (Fig. 4.12). The coated substrate was heat-treated at 180°C for 1 h in air, immersed further in a solution of *N*-dodecyl mercaptan (NDM, 0.25 mmol/L) in anhydrous ethanol for 24 h at room temperature, washed in ethanol, and dried at room temperature. The fabricated substrate exhibited excellent superhydrophobic (167.6°) and self-cleaning properties as well as robust mechanical abrasion and corrosion resistance.

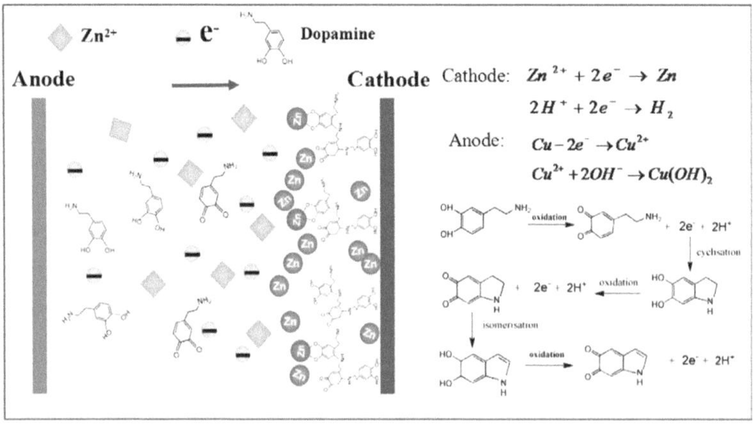

Figure 4.12 Schematic illustration of the deposition of electrolyte solution on copper substrate. Reprinted from Ref. [87]. Copyright (2016), with permission from Elsevier.

The surface of all the substrates showed similar CAs. The Zn ions were deposited uniformly on the substrates and the polymerization of DA occurred at the same time during electrochemical deposition. Moreover, continuous electrochemical deposition would decrease the pH of the electrolyte solution, which can modify the surface chemical composition of the anode substrate and strongly adhere to the substrate. The treated surface exhibited a more micronanohierarchical rough morphology after further modifying the coated substrate with an NDM solution, which imparted excellent superhydrophobicity to the substrates.

A hierarchical superhydrophobic pipeline steel (X90) substrate was fabricated by the electrochemical deposition of a Cu-Zn coating [88]. The pipeline steel plate was abraded mechanically using sandpaper, cleaned in acetone and ethanol, and surface-treated sequentially in an alkaline and acidic solution. Various Zn^{2+} to Cu^{2+} compositions were prepared by mixing the solution at molar ratios of copper sulfate pentahydrate ($CuSO_4 \cdot 5H_2O$) and zinc sulfate heptahydrate ($ZnSO_4 \cdot 7H_2O$) solutions of 0.2:1 to 2.5:1 followed by the addition of potassium sodium tartrate ($KNaC_4O_6H_4$, 100 g/L), and sodium hydroxide (NaOH, 50 g/L) solutions.

Electrochemical deposition was carried out by maintaining the pipeline steel at a fixed distance (20 mm) with various current densities (from 0.4 to 5.0 A/dm^2). The Cu-Zn ion-deposited steel substrate was modified by immersion in a potassium hydroxide (KOH, 2.5 mol/L) and ammonium persulfate ($(NH_4)_2S_2O_8$, 0.12 mol/L) solution to obtain a micronanohierarchical structure on the substrate. The substrate was modified further by immersion in a pentadecafluorooctanoic acid (90%, 0.01 mol/L) in ethanol solution for 12–72 h at room temperature followed by air-drying. The CA of the coated substrate was increased from hydrophilic to hydrophoboic and superhydrophobic by increasing the current density to 2 A/dm^2; the CA became stable (156°) by further increasing the current density to 5 A/dm^2 (Fig. 4.13). Moreover, the stable superhydrophobicity was obtained by immersing the substrate in a hydrophobic fluorine solution for 30 min.

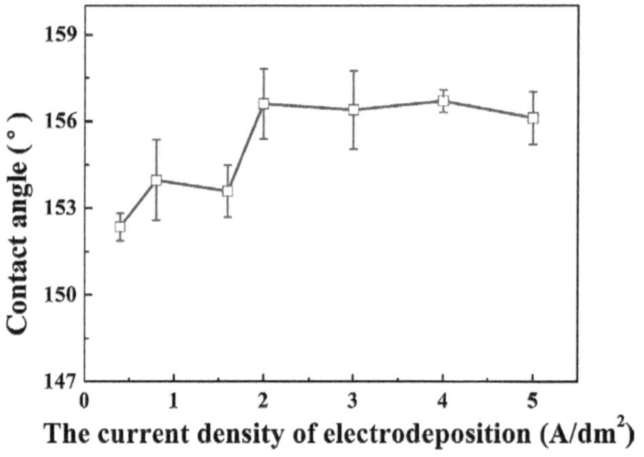

Figure 4.13 Contact angle by increasing the current density in electrodeposition process (from 0.4 to 5 A/dm^2). Reprinted from Ref. [88]. Copyright (2016), with permission from Elsevier.

A simple one-step electrodeposition process was also carried by the deposition of thin nickel stearate/nickel hydroxide films on an aluminum alloy substrate [89]. The fabricated aluminum alloy substrate exhibited superhydrophobicity due to the deposition of micronanoparticles of nickel ions and hydrophobic stearate molecules on the substrate. A cauliflower-like cluster was deposited on the aluminum alloy substrate at a lower Ni^{2+}/SA molar ratio in solution. The clusters size and formation of nanomicroparticles increased with increasing the Ni^{2+}/SA molar ratios in solution. The hydrophilicity (54° ± 2°) of the pristine precleaned substrate became hydrophobic (135° ± 1°) after deposition with a lower Ni^{2+}/SA molar ratio in solution, whereas the surface became superhydrophobic (160° ± 1°) after increasing the Ni^{2+}/SA molar ratio in solution during electrochemical deposition. A highly robust and porous superhydrophobic Ti6Al4V titanium alloy substrate was fabricated by surface treatment with fluoroalkylsilane (FAS) [90]. The fabricated substrate exhibited excellent anticorrosion behavior.

4.3.2 Physical Routes

Superhydrophobic organic-inorganic nanohybrid surfaces were fabricated by various physical approaches, such as the deposition of a thin layer of highly hydrophobic materials by spray, spin, and dip coating, as well as the formation of a layer-by-layer assembly of hydrophobic materials, nonwoven uniform nanofibers, and initial surface treatment using plasma, laser, and candle-soot, and posthydrophobic modification. The physical routes require generally two or more steps of surface modification for the fabrication of superhydrophobic coatings compared to the chemical approaches.

4.3.2.1 Spray coating

The fabrication of superhydrophobic coatings on various substrates was achieved by a spray coating method and the coated substrates were used in a wide range of applications. The superhydrophobic coatings were fabricated by spraying hydrophobic coating materials followed by subsequent curing. The hydrophobic coating materials were synthesized or prepared by either one or more steps, depending on the material conditions. The thickness and surface wettability of the coated substrate were controlled and tuned easily by adjusting the coating pressure, distance between the coating substrate and the spray gun, curing conditions, and coating materials.

Mahadik et al. fabricated superhydrophobic coatings on a glass substrate by spraying a hydrophobic solution prepared using the sol-gel technique [91]. A hydrophobic alcosol was prepared by the hydrolysis and condensation of MTES in methanol and aqueous ammonium hydroxide (13.6 M) as a catalyst and spray-coated on a precleaned glass substrate. A commercial spray-coating machine was used with a constant nozzle size diameter (0.5 mm), substrate-to-nozzle distance (30 cm), and pressure (90 psi). The alcosol was sprayed on the hot glass substrate (100°C) and annealed at 150°C for 2 h followed by subsequent calcination at 550°C for 1 h to obtain a uniform silica coating on the substrate. The calcined substrate was hydrophobic (94° ± 3°) with a high SA 39° ± 4°. The coated substrate was modified further by the deposition of a hydrophobic trimethylchlorosilane (TMCS) in hexane solution (10%). The TMCS-modified coated substrate was superhydrophobic (167° ± 1°) with a very low SA 3° ± 1° on the glass substrate. Moreover, the superhydrophobic glass substrate exhibited excellent transparency and thermally stable surface properties.

Cirisano et al. also used organic-inorganic hybrid materials for the fabrication of an amphiphobic coating on a glass substrate using the dispersion of a fumed silica nanoparticle/fluorinated polymer [92]. The hydrophobic dispersion was spray-coated on a glass substrate with a distance and pressure of 5 cm and 3 bar, respectively, and dried in an inert atmosphere. The coated substrate showed excellent repellence for water (170°) and various oils (paraffin oil, 160°; sunflower oil, 149°; and benzyl alcohol, 140°). Moreover, the coating materials exhibited stable amphiphobicity to simulated and real seawater environments as well as excellent antifouling properties.

Oligomer-wrapped silica particles was synthesized by a silanol–isocyanate reaction using dried silica nanoparticles (2.0 g) and tolylene diisocyanate (10.0 g) in the presence of triethylamine (30 ml) and dibutyltindilaurate (DBTDL, 1.0 wt%) catalyst [93]. The oligomer-wrapped silica nanoparticles were modified further with PDMS, 3-glycidoxypropyl trimethoxysilane (GPTMS), and 3-aminopropyl triethoxysilane using dibutyltindilaurate (DBTDL) in hexane and sprayed on a glass substrate. The fabricated glass substrate exhibited excellent superhydrophobic (166° ± 1°) property with a low SA (3°), self-recoverable, stress, and pH tolerance properties [93].

A superhydrophobic hybrid organic-inorganic hybrid coated substrate was developed by modifying the surface of alumina nanoparticles under reflection using a carboxylic acid-functionalized material and long-chain hydrophobic material (Scheme 4.4) [94]. The hydrophobic material was spray-coated on glass, cardboard, or paper substrate and dried at either 25°C or 40°C. The fabricated substrate was superhydrophobic.

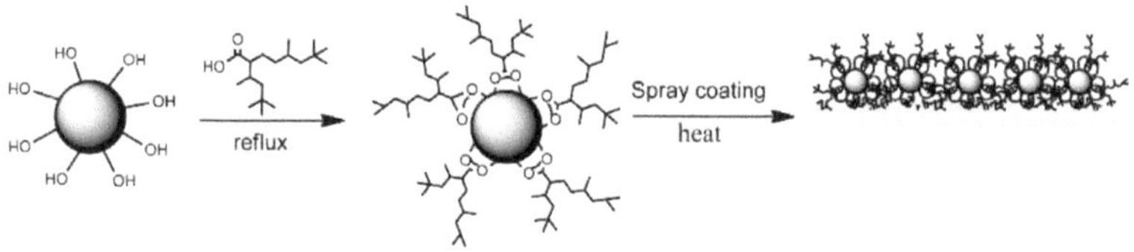

Scheme 4.4 Schematic representation of fabrication of superhydrophobic surfaces. Reprinted from Ref. [94]. Copyright (2016), with permission from the American Chemical Society.

Wang et al. prepared superhydrophobic PDMS surfaces by modifying with ZnO and carbon nanotubes (CNTs). The prepared ZnO-PDMS and CNT-PDMS were mixed with polyphenylene sulfide (PPS) and sprayed on a glass substrate to obtain the superhydrophobic surfaces [95]. The superhydrophobic surfaces exhibited excellent pH and heat resistance properties.

Recently, colorful superhydrophobic coatings were also fabricated on a steel substrate by spraying (spray gun with nozzle diameter 0.8 mm, pressure 0.2 MPa, and distance 15 cm) a mixed solution of various inorganic salts (cupric acetate, iron(III) chloride hexahydrate, cobaltous(II) sulfate heptahydrate, chromium(III) chloride hexahydrate, and zinc chloride, individually) with a sodium stearate (NaSA) suspension in ethanol [96]. The coated substrate exhibited superhydrophobicity with various colors (such as blue, aurantium, purple, cinerous, and white, respectively) according to the type of inorganic salt used in the mixing composition. Moreover, the fabricated substrate showed excellent anticorrosion properties, which was confirmed by immersing the coated substrate in a 3.5% NaCl solution.

A transparent superhydrophobic paper was fabricated by modifying surface of silica nanoparticles with dodecyltrichlorosilane followed by spraying on the standard printer-grade paper [97]. The coated paper substrate exhibited excellent transparency and superhydrophobicity (Fig. 4.14). The letters in the printed and hydrophobic coated paper are clearly visible to the naked eye, underscoring the higher transparency of the coating material. Moreover, the coated superhydrophobic paper was highly flexible and maintained the surface properties on the substrate, even after folding or pressing (CA and SA of the paper before pressing were 155° and 7.2°, respectively, and after pressing were 153° and 9.5°, respectively) mechanically several times. Similar superhydrophobicity was also obtained by spraying on the cotton fabrics. Several hydrophobic materials were prepared by simple mixing followed by subsequent spraying on various substrates and superhydrophobicity were obtained on the substrates.

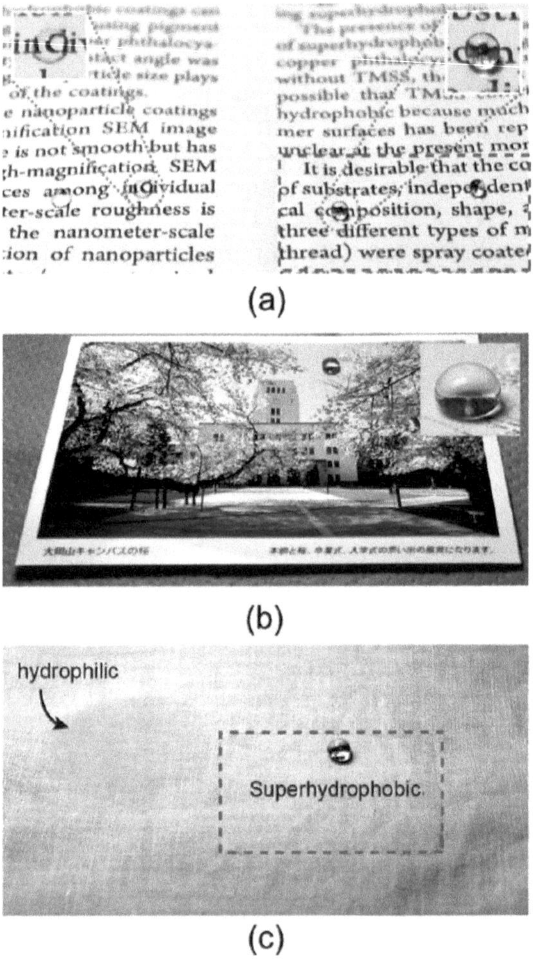

Figure 4.14 Optical image of the surface wettability on (a) paper, (b) printed photograph, and (c) cotton (uncoated and coated parts show hydrophilic and superhydrophobic property). Reprinted from Ref. [97]. Copyright (2012), with permission from the American Chemical Society.

4.3.2.2 Spin coating

Superhydrophobic surfaces were developed easily by spin coating with the deposition and rotation of highly hydrophobic organic-inorganic hybrid solutions followed by mild or strong thermal curing in a conventional drying oven. The transparency of the substrate can also be controlled using this coating method. A novel superhydrophobic surface was fabricated with excellent transparency on the substrate by a simple one-step hydrophobic surface modification of silica nanoparticles followed by spin coating on the substrate [98].

The fabricated substrate showed a micronanohierarchical surface morphology due to the reaction of highly hydrophobic long-alkyl-chain fluorosilane (heptadecafluoro-1,1,2,2,-tetrahydrodecyl) dimethylchlorosilane (HDFTHD) (99%) in hydrophilic silica nanoparticles in the presence of trimethylamine in toluene. A Si wafer (P type, polished) substrate that had been pretreated by surface modification with 3-(triethoxysilyl) propyl succinic anhydride (TESPSA) (95%) was spin coated. The fluorosilane-modified silica nanoparticles (F-SiO_2 NP) were dispersed in fluorinated solvent (Novec 7300) and spin-coated (1500 rpm speed for 20 s at a velocity of 150 rpm/s) on a TESPSA surface-treated Si wafer substrate. The spin-coated substrate exhibited superhydrophobicity and excellent transparency (95% transmittance in the visible region) (Fig. 4.15).

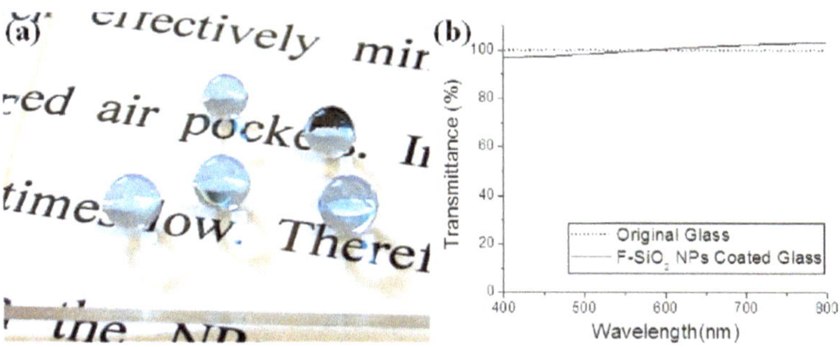

Figure 4.15 Optical transparency of spin-coated F-SiO_2 NP film (1.0 wt %) on a glass substrate. (a) Photograph of water droplets on F-SiO_2 nanoparticle–coated glass substrate. (b) UV-vis spectrum of the glass substrates with and without F-SiO_2 NP coating. A small amount of dimethyl methylene blue dye was dissolved in water for illustration purpose. Reprinted from Ref. [98]. Copyright (2012), with permission from the American Chemical Society.

Moreover, the surface wettability on the coated substrates were controlled and improved by increasing the concentration of hydrophobic fluorinated silica nanoparticles in the dispersion. The self-assembly of fluorinated silica particles was increased on the coated substrate, leading to a denser coating at higher concentrations, which may eventually enhance the surface roughness and produce a hierarchical surface morphology. The hierarchical surface morphology with hydrophobic fluoro-silane on the surface can repel the water droplets on the substrate.

A superhydrophobic surface was fabricated by spin-coating a solution of Teflon AF1600 (Dupont, U.S.A.) and FC-770 fluorinert (3M, U.S.A.) liquid on a copper substrate [99]. The pristine copper substrate was hydrophobic with a CA of 108.44° ± 4.68°. In contrast, the pre-cleaned copper substrate by chemical polishing with nitric acid (HNO_3), followed by oxidizing with hydrogen peroxide (H_2O_2), was superhydrophilic due to the formation of cupric oxide (CuO) on the surface of the copper substrate, which has strong affinity to adhere with water due to the hydrophilic nature of CuO. The superhydrophilic substrate becomes superhydrophobic by simple one-step spin coating (speed and spin time were 4000 rpm and 30 s after prespinning, respectively) with a mixture of hydrophobic Teflon and FC-770 fluorinert liquid followed by heat treatment on a hot plate at 60°C for 60 min. The fabricated copper substrate exhibited superhydrophobicity with a CA of 151.69° ± 5.11°.

Söz et al. prepared a PS/silica dispersion and bisphenol-A-based epoxy resin (ER)/silica dispersions for the fabrication of a superhydrophobic surface on a glass substrate [100]. Hydrophobic fumed silica HDK H2000 (H2K) particles (5–30 nm) were dispersed vigorously with PS and spin-coated on a glass substrate and dried overnight at room temperature and then at 40°C for 24 h. Similarly, ER/silica dispersions were prepared by mixing hydrophobic fumed silica particles with epoxy resin, diamine hardener (2-methyl-1,5-diaminopentane), and polyperfluoroether diol (Fluorolink E10-H) (PFE) or polycaprolactone-polydimethylsiloxane-polycaprolactone (PCL-PDMS-PCL) triblock copolymer (PCL and PDMS block lengths of 2000 and 3000 g/mol, respectively; 2-3-2). The resulting dispersions were then spin-coated on a glass substrate and dried overnight at room temperature and then at 150°C for 5 h. The substrates fabricated from either PS/silica or ER/silica dispersions exhibited superhydrophobicity due to the agglomeration of hydrophobic silica nanoparticles with PS or ER, which enhances the surface roughness and hierarchical surface morphologies.

4.3.2.3 Drop coating

Drop casting is the simplest way of forming any surface property on a substrate by simply dropping a hydrophilic or hydrophobic solution followed by manual or automatically casting on the substrate. On the other hand, the surface properties depend on the material to be coated on the substrate. The thickness of the coated substrate can be controlled on the basis of the solution concentration. In general, organic solvents, such as ethanol, toluene, tetrahydrofuran, hexane, and halogenated solvents, are used to solubilize or disperse the hydrophilic or hydrophobic materials.

A better uniform coating can be achieved using organic solvents due to the easy film forming ability on a substrate than the use of water as a solvent. The choice of organic solvents also control the film thickness based on the rate solvent evaporation from the coated substrate. On the other hand, drop casting can result in an uneven thickness of the coated film when there are fluctuations of the concentration and evaporation rate of organic solvents on the coated substrate. Recently, superhydrophobic surfaces were also fabricated on various substrates by drop casting followed by room-temperature curing or mild thermal treatment.

Nagappan et al. prepared an environmentally friendly organic-inorganic hybrid material dispersion in a few steps followed by casting on various substrates. The hybrid suspension was prepared from lotus leaf powder (LLP), which had been prepared by the washing and drying of fresh lotus leaves followed by grinding into a fine powder and sieving through a sieve mesh. The superhydrophobic polysiloxane powder was synthesized by the synthesis of phenyl-substituted silica ormosil (PSiOr) suspension [101]. A desirable amount of superhydrophobic polysiloxane powder and LLP was dispersed in ethanol followed by the addition of PSiOr suspension with stirring can lead to the formation of hierarchical micronanoparticles [101].

Drop casting of the resulting suspension can result in the evaporation of solvents much faster from the surface and produce excellent and stable superhydrophobicity on a substrate [101]. The formation of a micronanohierarchical surface morphology with higher roughness on the coated substrate is responsible for repelling the water droplet from the substrate. The hybrid (LLPPSiOr)-coated glass substrate exhibited good thermal stability, and the superhydrophobicity was maintained by annealing the coated substrate at 500°C for 8 h; the surface was rendered superhydrophilic by annealing at 500°C for more than 8 h, whereas the surface became superhydrophilic faster at 550°C and 600°C (Fig. 4.16) [101]. The change from superhydrophobicity to superhydrophilicity was due to the decomposition of the organic functional groups at the surface of the hybrid material.

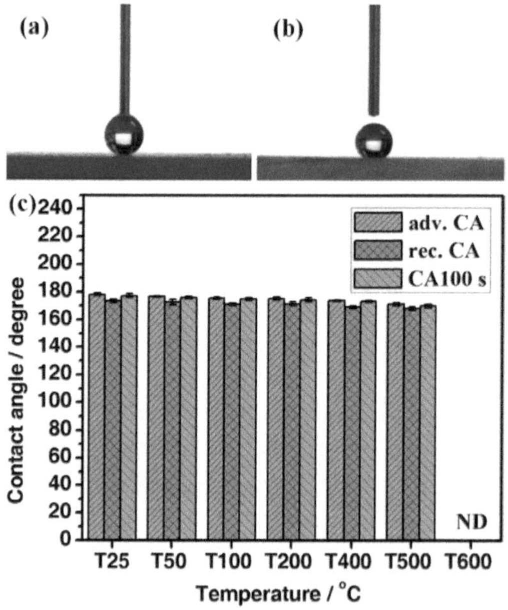

Figure 4.16 (a and b) The highly stable superhydrophobic surface maintained superhydrophobic properties up to 500°C for 5 h and more than 6000 s upon contact with water. (c) DCA of the cast hybrid substrate cured at various temperatures for 5 h by slow heating (1 h to the desired temperature) and cooling to room temperature. adv. CA, advancing contact angle; rec. CA, receding contact angle; T, temperature; s, seconds; ND, not detected. Reprinted from Ref. [101] with permission from The Royal Society of Chemistry.

Moreover, the coated substrate exhibited excellent non-adhesion behavior to water droplets. The dropped water droplet could pass away easily from a superhydrophobic surface (Fig. 4.17). The superhydrophobicity was also obtained on a substrate by spin coating, where the surface property was measured according to the rotation speed of the suspension during spin coating [102].

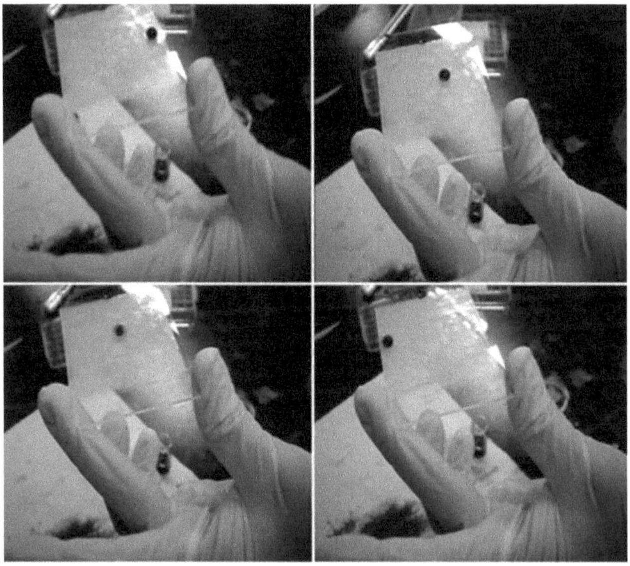

Figure 4.17 Optical images of non-stick coatings prepared by spin coating at 1000 rpm on a glass-supported laminating film substrate. Reprinted from Ref. [102]. Copyright (2014), with permission from Taylor & Francis.

The results showed similar surface superhydrophobicity on the substrate by drop casting and spin coating of the organic-inorganic hybrid suspension [101, 102]. This simple approach can also produce superhydrophobicity on a range of substrates (Fig. 4.18) [103]. Furthermore, the surface property of the hybrid suspension was checked by examining the effects of hydrophilic silica nanoparticles or silica ormosils instead of a PSiOr suspension [104].

A stable hydrophobic surface was formed on a glass substrate. The effects of tree leaves on the fabrication of the superhydrophobic surface on glass and other substrates were also examined [103, 105, 106]. They also used *Ailanthus altissima, Camellia japonica,* and *Pisdium gujava* leaf powder instead of LLP to produce superhydrophobic hybrid materials and surfaces.

The prepared hybrid suspensions coated on a glass substrate exhibited superhydrophobicity as well as nonstick and self-cleaning properties (Fig. 4.18). The superhydrophobic organic-inorganic hybrid substrate also showed excellent thermal stability as well as cold and hot water resistance (Fig. 4.19). The superhydrophobic powder was also obtained by evaporating and washing the hybrid powder. The prepared superhydrophobic powder showed the selective adsorption of metal ions in the mixed-metal ion solution made from artificial seawater [105].

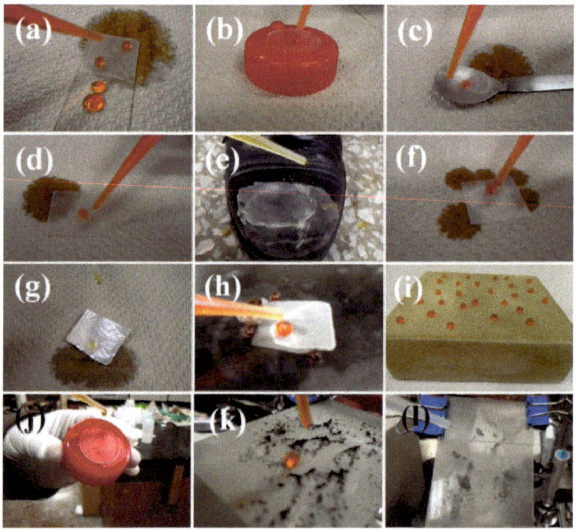

Figure 4.18 Superhydrophobicity on a wide range of substrates. (a–h) LLPPSiOr hybrid micronanocomposites suspension coated on a glass, plastic cap, metal spatula, laminating film and shoe, weighing paper, aluminum foil, and wooden board. (i) Large-scale fabrication of a dip-coated LLPPSiOr hybrid melamine sponge (water colored with methyl orange for visual purposes). (j–l) The nonstick and self-cleaning superhydrophobic hybrid substrates at the sliding angle. Reprinted from Ref. [103]. Copyright (2014), with permission from Springer.

A superhydrophobic surface was also obtained by various other researchers by the simple drop casting of a hydrophobic solution on glass or other substrates and used for a range of applications [107–109]. Silica nanoparticles with a controlled particle size were synthesized, and their surface properties were modified with multifunctional groups. The prepared multifunctional dispersion was drop-casted onto a glass substrate and the surface properties were examined. The fabricated organic-inorganic hybrid coated substrate exhibited highly robust superhydrophobicity and self-cleaning properties [107]. CNTs modified with PDMS and SiO_2 nanoparticles were used to fabricate a superhydrophobic coating by a drop-casting method on a glass substrate. The superhydrophobic glass substrate exhibited excellent stability, a sheet resistance of 13.9 kΩ sq^{-1}, transparency of more than 80% to visible light, and stable surface properties against UV irradiation [109].

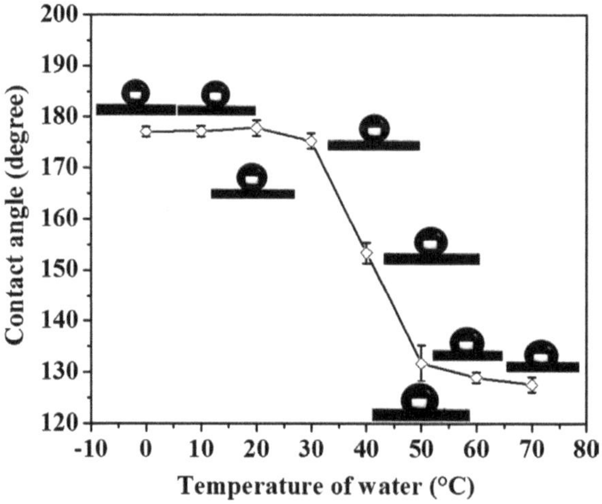

Figure 4.19 Effect of water droplet temperatures on the superhydrophobic surface. Inset: Optical images illustrate the CA images of the water droplet. Reprinted from Ref. [103]. Copyright (2014), with permission from Springer.

4.3.2.4 Electrospinning

Electrospinning has attracted considerable attention for the fabrication of superhydrophobic nanofibrous mat membranes. Sas et al. briefly reviewed the fabrication of superhydrophobic and self-cleaning surfaces with various polymers by electrospinning and explained various experimental conditions to develop a superhydrophobic and self-cleaning nanofibrous mat [110]. In most cases, hydrophobic polymers were used to fabricate superhydrophobic surfaces on the nanofibrous mat due to the formation of hydrophobicity on the substrate, 3D porous nonwoven web structure, and air pockets at the porous structure, which eventually resist the adhesion of water droplets on the surface. In some cases, hydrophilic polymers were also used to fabricate superhydrophobic surfaces, but this may require additional surface treatment with hydrophobic materials. The resulting superhydrophobic nanofibrous web exhibited excellent water repellence and self-cleaning properties due to the presence of high roughness and LSEMs on the surface of the nanofibrous web.

A nanofibrous mat was developed using a colloidal electrospinning technique for membrane distillation (MD) [111]. A hydrophobic silica nanoparticle dispersion was prepared by modifying the surface properties of silica nanoparticles (40 nm) with a hydrophobic silane coupling agent or surface modifier (octadecyltrichlorosilane, OTS). A colloidal solution was prepared by dispersing the OTS-modified silica nanoparticles (18 wt%) in DMF. Similarly, a polyvinylidene (PVDF, 18 wt%) solution was also prepared in DMF. The OTS-silica nanoparticles/DMF dispersion was added to the PVDF/DMF dispersion at a 1:2 ratio and mixed strongly to produce a homogenized dispersion that was used for the fabrication of a nanofibrous scaffold. The nanofibrous scaffold showed a micronanohierarchical surface morphology and superhydrophobicity due to the combination of inorganic nanoparticles and organic PVDF.

The surface properties of the nanofibrous scaffold were also tuned by varying the silica nanoparticle size (147, 210 nm). The PVDF scaffolds exhibited hydrophobicity (135.5° ± 1.8°), which became superhydrophobic after mixing with OTS-modified silica nanoparticles. The superhydrophobic surface CAs increased with increasing silica nanoparticle size (152.3° ± 2.0°, 155.6° ± 2.3°, and 163.1° ± 1.9°). The results suggest that controlled surface wettability can be developed by changing the particle size of the nanoparticles followed by hydrophobic surface treatment. Moreover, the resulting scaffold exhibited excellent superhydrophobicity in MD applications.

Shami et al. fabricated an organic-inorganic hybrid nanofabric by electrospinning using organic porous poly(acrylonitrile) (PAN) and inorganic MgAl-layered double hydroxide (LDH) (Fig. 4.20) [112]. The nanofabric was obtained by first preparing a nanomembrane by electrospinning using PAN and polymethylmethacrylate (PMMA). A porous PAN nanomembrane was obtained by removing the PMMA in the membrane using an acetic acid solution. The Mg-Al LDH was loaded into the porous PAN nanomembrane by a hydrothermal method and surface-treated with cyclohexanecarboxylic acid. The resulting nanofabric exhibited a hierarchical and 3D porous surface morphology as well as excellent superhydrophobic stability against cold (0°C) and hot (95°C) water. The porous nanofabric also exhibited excellent stability against the separation of various oils and organic solvents in water with a separation efficiency of over 98.4%.

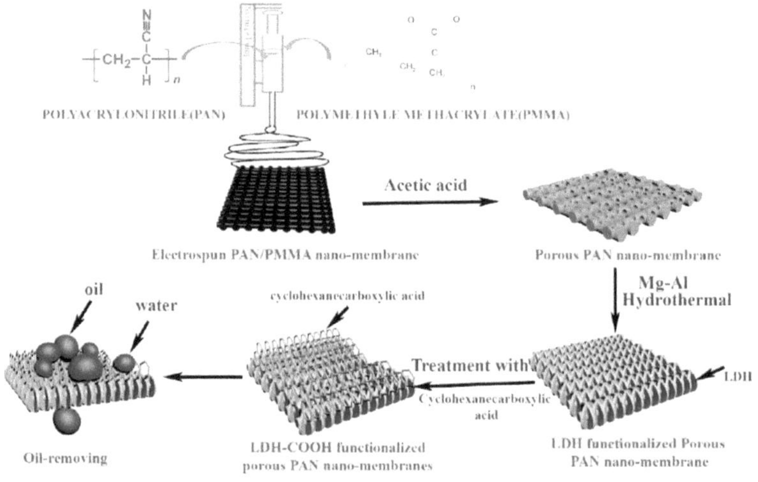

Figure 4.20 Schematic illustration of 3D hierarchical roughness MgAl–LDH-functionalized electrospun porous PAN nanomembranes. Reprinted from Ref. [112]. Copyright (2016), with permission from the American Chemical Society.

A superhydrophobic organic-inorganic hybrid nanofibrous mat was also fabricated using cellulose acetate nanofibers and modified with 1H,1H,2H,2H perfluorooctyltriethoxysilane (FS) via a hydrolysis and condensation reaction [113]. The superhydrophobic hybrid mat showed excellent separation efficiency for oils and organic solvents in water. Dong et al. prepared a superhydrophobic/hydrophobic nanofibrous surface using natural protein (zein) extracted from corn endosperm [114]. The zein-casting film (ZCF) was hydrophilic (76.5° ± 1.1°), whereas the zein electrospun nanofibrous network (ZENN) was hydrophobic to superhydrophobic according to the concentration of the zein concentration. Lowering the concentration of zein in an ethanol solution (5 wt%) resulted in higher hydrophobicity (130°), whereas the surface became superhydrophobic at 10 wt% and 15 wt% (153.6° ± 2.1° and 150.1° ± 1.3°, respectively). Further increasing the concentration of zein in the solution resulted in a decrease in the superhydrophobicity of the fabricated nanofibrous that eventually became hydrophobic (143.8° ± 2.4°, 141.0° ± 1.6°, and 136.4° ± 2.1°) by (20 wt%, 25 wt%, and 30 wt%, respectively).

The decrease in the surface wettability was attributed to the formation of various surface morphologies ranging from beads to, beads/fibers, and fibers. The fabricated surface showed different cell adhesion behaviors, depending on the surface wettability. A higher surface contact angle (CA) would show better cell adhesion to the surface. A nanofibrous mat produced by the simple mixing of crosslinked PDMS and fluorodecyl POSS (50 wt%) showed excellent superomniphobicity (repelling both aqueous and organic solvents) [115]. Several other studies examined the fabrication of superhydrophobic organic-inorganic nanofibrous mats and their applications are emerging in many fields.

4.3.2.5 Plasma treatment

Plasma treatment is used widely to fabricate surfaces with a hierarchical and controlled morphology. Moreover, the surface properties of a substrate can be altered to either hydrophilic or hydrophobic depending on the type of reacting gases used (O_2, Ar, CH_4, and C_4H_8). In general, plasma light generates electrons, active ions, and radicals from polymers and various substrates, which emphasizes the surface wettability or further modification of the material. Three different approaches (i.e., plasma etching, plasma polymerization, and plasma sputtering) were used to modify the surface wettability of a substrate. In most cases, plasma etching was used to fabricate a superhydrophobic coating due to the easier surface modification and making higher roughness on the substrate than the other two cases, which was further treated with a hydrophobic material to render it superhydrophobic. The surface wettability also depends on the plasma power source and the plasma treatment time.

Kumar et al. fabricated a superhydrophobic surface by plasma treatment on a p-type silicon wafer (100) substrate [116]. The silicon wafer was precleaned in a mixture of sulfuric acid

(H_2SO_4) (98%) and H_2O_2 (30%) solution at a 3:1 ratio for 15 min, washed with deionized water, and dried with nitrogen gas. A SU8 (Microchem, SU8-2002) negative photoresist was spin-coated on the silicon wafer surface, followed by UV exposure and plasma treatment on the substrate using O_2, SF_6, and a combination of both O_2 and SF_6 plasma treatment. The self-assembled monolayer (SAM) of octadecyltrichlorosilane (OTS) was achieved by immersing the silicon substrate in the OTS solution (Fig. 4.21). The OTS-modified substrate exhibited excellent superhydrophobicity when using a SU8 photoresist followed by the above plasma treatments (O_2, SF_6, and a combination of both O_2 and SF_6), whereas the OTS-modified silicon substrate was hydrophilic or hydrophobic without the use of a SU8 photoresist and plasma treatments.

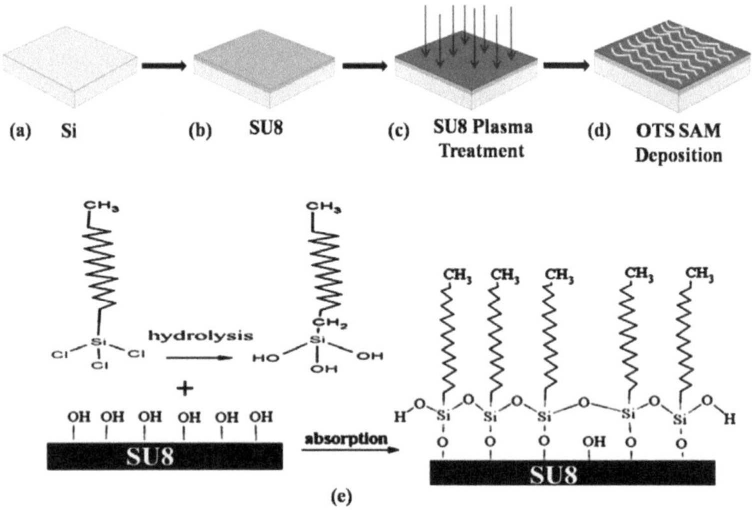

Figure 4.21 Schematic process flow for OTS SAM deposition on plasma-treated surface: (a) cleaned Si wafer, (b) SU8 spinning, (c) plasma treatment, (d) OTS SAM depositions, and (e) formation mechanism of OTS SAM on SU8 surface. Reprinted from Ref. [116]. Copyright (2015), with permission from WILEY-VCH Verlag GmbH.

The self-healing superhydrophobic fabrics were obtained by surface modification of the fabrics when immersed in polydopamine nanocapsule-encapsulated hydrophobic agents, such as polydopamine@octadecylamine (PDA@ODA) or polydopamine@

octadecanethiol (PDA@ ODT) [117]. The resulting superhydrophobic surfaces showed excellent stability against several washing cycles and they were flexible. The superhydrophobic (150°) fabric surface became superhydrophilic (0°) after surface treatment with oxygen (O_2) plasma due to the generation of active oxygen radicals at the surface, which strongly react with water droplet and penetrates the surface. On the other hand, the superhydrophilic surface recovered slowly to a superhydrophobic one by increasing the stretching cycles (2500). The authors also prepared a superhydrophobic (152°) sponge by modification with PDA@ODT. The PDA@ODT-coated sponge became superhydrophobic (152°) after the plasma treatment, which recovered slowly by compression of the sponge more than 250 times.

Li et al. developed a superhydrophobic surface on a stainless steel substrate by a radio-frequency (RF) plasma coating [118]. The stainless steel substrate was etched in a strong acid solution followed by drying and deposition of a mixture of pentafluoroethane precursor (Praxair, flow rate 20 SCCM, standard cubic centimeters per minute) and argon (75 SCCM) at 110°C and 120 W. The highly crosslinked fluoropolymer was deposited uniformly and bonded covalently to the stainless steel substrate with a thickness of approximately 100 nm. When the fluoropolymer was deposited on the chemically etched stainless steel substrate, the substrate surface changed from hydrophobic to superhydrophobic according to the duration and type of chemical etching prior to hydrophobic surface treatment.

Recently, glow discharge electrolysis plasma (GDEP) has become interesting for the fabrication of environmentally friendly superhydrophobic organic-inorganic coatings on a zinc substrate [119]. The superhydrophobic surface was obtained by precleaning the zinc sheet in acetone and ethanol followed by etching and functionalization in the GDEP reactor. After the initial etching, the zinc substrate was treated further by immersion in a hydrophobic stearic acid ethanol solution at 25°C for approximately 12 or 24 h via GDEP and rinsed with ethanol and distilled water, and dried at room temperature.

References

1. Bhushan, B., Jung, Y. C., and Koch, K. (2009). Self-cleaning efficiency of artificial superhydrophobic surfaces, *Langmuir*, **25**, pp. 3240–3248.
2. Cao, L., Jones, A. K., Sikka, V. K., Wu, J., and Gao, D. (2009). Anti-icing superhydrophobic coatings, *Langmuir*, **25**, pp. 12444–12448.
3. Khalil-Abad, M. S., and Yazdanshenas, M. E. (2010). Superhydrophobic antibacterial cotton textiles, *J. Colloid Interface Sci.*, **351**, pp. 293–298.
4. Zhou, G., He, J., Gao, L., Ren, T., and Li, T. (2013). Superhydrophobic self-cleaning antireflective coatings on Fresnel lenses by integrating hydrophilic solid and hydrophobic hollow silica nanoparticles, *RSC. Adv.*, **3**, pp. 21789–21796.
5. Wang, J., Zheng, Y., and Wang, A. (2012). Superhydrophobic kapok fiber oil-absorbent: preparation and high oil absorbency, *Chem. Eng. J.*, **213**, pp. 1–7.
6. Zhang, X., Jin, M., Liu, Z., Tryk, D. A., Nishimoto, S., Murakami, T., and Fujishima, A. (2007). Superhydrophobic TiO_2 surfaces: preparation, photocatalytic wettability conversion, and superhydrophobic–superhydrophilic patterning, *J. Phys. Chem. C*, **111**, pp. 14521–14529.
7. Mehmood, U., Al-Sulaiman, F. A., Yilbas, B. S., Salhi, B., Ahmed, S. H. A., and Hossain, M. K. (2016). Superhydrophobic surfaces with antireflection properties for solar applications: a critical review, *Solar Energy Mater. Solar Cells*, **157**, pp. 604–623.
8. Nagappan, S., Park, S. S., and Ha, C. S. (2015). Recent advances in superhydrophobic nanomaterials and nanoscale systems, *J. Nanosci. Nanotechnol.*, **14**, pp. 1441–1462.
9. Nagappan, S., and Ha, C. S. (2015). Emerging trends in superhydrophobic surface based magnetic materials: fabrications and their potential applications, *J. Mater. Chem. A*, **3**, pp. 3224–3251.
10. Han, Z., Mu, Z., Yin, W., Li, W., Niu, S., Zhang, J., and Ren, L. (2016). Biomimetic multifunctional surfaces inspired from animals, *Adv. Colloid Interface Sci.*, **234**, pp. 27–50.
11. Wen, L., Tian, Y., and Jiang, L. (2015). Bioinspired super-wettability from fundamental research to practical applications, *Angew. Chem. Int. Ed.*, **54**, pp. 3387–3399.
12. Kuang, M., Wang, J., and Jiang, L. (2016). Bio-inspired photonic crystals with superwettability, *Chem. Soc. Rev.*, **45**, pp. 6833–6854.

13. Cohen, N., Dotan, A., Dodiuk, H., and Kenig, S. (2016). Superhydrophobic coatings and their durability, *Mater. Manuf. Processes*, **31**, pp. 1143–1155.

14. Simpson, J. T., Hunter, S. R., and Aytug, T. (2015). Superhydrophobic materials and coatings: a review, *Rep. Prog. Phys.*, **78**, pp. 086501 (1–14).

15. Li, L., Li, B., Dong, J., and Zhang, J. (2016). Roles of silanes and silicones in forming superhydrophobic and superoleophobic materials, *J. Mater. Chem. A*, **4**, pp. 13677–13725.

16. Lai, Y., Huang, J., Cui, Z., Ge, M., Zhang, K., Chen, Z., and Chi, L. (2016). Recent advances in TiO_2 -based nanostructured surfaces with controllable wettability and adhesion, *Small*, **12**, pp. 2203–2224.

17. Milionis, A., Bayer, I. S., and Loth, E. (2016). Recent advances in oil-repellent surfaces, *Int. Mater. Rev.*, **61**, pp. 101–126.

18. Wang, Z., Elimelech, M., and Lin, S. (2016). Environmental applications of interfacial materials with special wettability, *Environ. Sci. Technol.*, **50**, pp. 2132–2150.

19. Khojasteha, D., Kazeroonib, M., Salariana, S., and Kamali, R. (2016). Droplet impact on superhydrophobic surfaces: a review of recent developments, *J. Ind. Eng. Chem.*, **42**, pp. 1–14.

20. Ionov, L., and Synytska, A. (2012). Self-healing superhydrophobic materials, *Phys. Chem. Chem. Phys.*, **14**, pp. 10497–10502.

21. Li, Y., Li, L., and Sun, J. (2010). Bioinspired self-healing superhydrophobic coatings, *Angew. Chem. Int. Ed.*, **49**, pp. 6129–6133.

22. Chen, K., Zhou, S., Yang, S., and Wu, L. (2015). Fabrication of all-water-based self-repairing superhydrophobic coatings based on UV-responsive microcapsules, *Adv. Fun. Mater.*, **25**, pp. 1035–1041.

23. Chen, K., Gu, K., Qiang, S., and Wang, C. (2017). Environmental stimuli-responsive self-repairing water based superhydrophobic coatings, *RSC Adv.*, **7**, pp. 543–550.

24. Lowe, A. B. (2014). Thiol–ene "click" reactions and recent applications in polymer and materials synthesis: a first update, *Polym. Chem.*, **5**, pp. 4820–4870.

25. Fang, Y., Ha, H., Shanmuganathan, K., and Ellison, C. J. (2016). Polyhedral oligomeric silsesquioxane-containing thiol–ene fibers with tunable thermal and mechanical properties, *ACS Appl. Mater. Interfaces*, **8**, pp. 11050–11059.

26. Chen, S., Li, X., Li, Y., and Sun, J. (2015). Intumescent flame-retardant and self-healing superhydrophobic coatings on cotton fabric, *ACS Nano*, **9**, pp. 4070–4076.

27. Chen, Q., de Leon, A., and Advincula, R. C. (2015). Inorganic-organic thiol–ene coated mesh for oil/water separation, *ACS Appl. Mater. Interfaces*, **7**, pp. 18566–18573.

28. Sparks, B. J., Hoff, E. F. T., Xiong, L., Goetz, J. T., and Patton, D. L. (2013). Superhydrophobic hybrid inorganic-organic thiol-ene surfaces fabricated via spray-deposition and photopolymerization, *ACS Appl. Mater. Interfaces*, **5**, pp. 1811–1817.

29. Xiong, L., Kendrick, L. L., Heusser, H., Webb, J. C., Sparks, B. J., Goetz, J. T., Guo, W., Stafford, C. M., Blanton, M. D., Nazarenko, S., and Patton, D. L. (2014). Spray-deposition and photopolymerization of organic-inorganic thiol–ene resins for fabrication of superamphiphobic surfaces, *ACS Appl. Mater. Interfaces*, **6**, pp. 10763–10774.

30. Dyett, B. P., Wu, A. H., and Lamb, R. N. (2014). Toward superhydrophobic and durable coatings: effect of needle vs crater surface architecture, *ACS Appl. Mater. Interfaces*, **6**, pp. 9503–9507.

31. Cho, Y. S., Choi, S. Y., Kim, Y. K., and Yi, G. R. (2012). Bulk synthesis of ordered macroporous silica particles for superhydrophobic coatings, *J. Colloid Interface Sci.*, **386**, pp. 88–98.

32. Ma, J., Zhang, X., Bao, Y., and Liu, J. (2015). Facile spraying method for fabricating superhydrophobic leather coating, *Colloids Surf. A: Physicochem. Eng. Aspects*, **472**, pp. 21–25.

33. Li, K., Zeng, X., Li, H., and Lai, X. (2014). Fabrication and characterization of stable superhydrophobic fluorinated-polyacrylate/silica hybrid coating, *Appl. Surf. Sci.*, **298**, pp. 214–220.

34. Chakradhar, R. P. S., Dinesh Kumar, V., Shivakumara, C., Raoc, J. L., and Basu, B. J. (2012). Characterisation of microstructure and evaluation of optical and EPR properties of superhydrophobic copper dodecanoate films, *Surf. Interface Anal.*, **44**, pp. 412–417.

35. Zhang, Y. Y., Ge, Q., Yang, L. L., Shia, X. J., Li, J. J., Yang, D. Q., and Sacher E. (2015). Durable superhydrophobic PTFE films through the introduction of micro- and nanostructured pores, *Appl. Surf. Sci.*, **339**, pp. 151–157.

36. Chen, K., Zhou, S., Yang, S., and Wu, L. (2014). Fabrication of all-water-based self-repairing superhydrophobic coatings based on UV-responsive microcapsules, *Adv. Funct. Mater.*, **25**, pp. 1035–1041.

37. Chen, K., Gu, K., Qiang, S., and Wang, C. (2017). Environmental stimuli-responsive self-repairing water based superhydrophobic coatings, *RSC Adv.*, **7**, pp. 543–550.

38. Tian, H., Yang, T., and Chen, Y. (2009). Fabrication and characterization of superhydrophobic thin films based on TEOS/RF hybrid, *Appl. Surf. Sci.*, **255**, pp. 4289–4292.

39. Xu, D., Wang, M., Ge, X., Lam, M. H. W., and Ge, X. (2012). Fabrication of raspberry SiO_2/polystyrene particles and superhydrophobic particulate film with high adhesive force, *J. Mater. Chem.*, **22**, pp. 5784–5791.

40. Jiang, W., Grozea, C. M., Shi, Z., and Liu, G. (2014). Fluorinated raspberry-like polymer particles for superamphiphobic coatings, *ACS Appl. Mater. Interfaces*, **6**, pp. 2629–2638.

41. Bayer, I. S., Steele, A., Martorana, P. J., and Loth, E. (2010). Fabrication of superhydrophobic polyurethane/organoclay nano-structured composites from cyclomethicone-in-water emulsions, *Appl. Surf. Sci.*, **257**, pp. 823–826.

42. Wang, W., Lockwood, K., Boyd, L. M., Davidson, M. D., Movafaghi, S., Vahabi, H., Khetani, S. R., and Kota, A. K. (2016). Superhydrophobic coatings with edible materials, *ACS Appl. Mater. Interfaces*, **8**, pp. 18664–18668.

43. Dong, F., Xie, H., Zheng, Q., and Ha, C. S. (2017). Superhydrophobic polysilsesquioxane/polystyrene microspheres with controllable morphology: from raspberry-like to flower-like structure, *RSC Adv.*, **7**, pp. 6685–6690.

44. Qin, L., Zhao, J., Lei, S., and Pan, Q. (2014). A smart "strider" can float on both water and oils, *ACS Appl. Mater. Interfaces*, **6**, pp. 21355–21362.

45. Alexander, S., Eastoe, J., Lord, A. M., Guittard, F., and Barron, A. R. (2016). Branched hydrocarbon low surface energy materials for superhydrophobic nanoparticle derived surfaces, *ACS Appl. Mater. Interfaces*, **8**, pp. 660–666.

46. Gao, Q., Hu, J., Li, R., Pang, L., Xing, Z., Xu, L., Wang, M., Guo, X., and Wu, G. (2016). Preparation and characterization of superhydrophobic organic-inorganic hybrid cotton fabrics via γ-radiation-induced graft polymerization, *Carbohydr. Polym.*, **149**, pp. 308–316.

47. Dong, Z. Q., Wang, B. J., Ma, X. H. Wei, Y. M., and Xu, Z. L. (2015). FAS grafted electrospun poly(vinyl alcohol) nanofiber membranes with robust superhydrophobicity for membrane distillation, *ACS Appl. Mater. Interfaces*, **7**, pp. 22652–22659.

48. Zhang, J., Cai, J., and Li, M. (2016). Grafting of PMMA brushes layer on Cu surface to create a stable superhydrophobic surface, *Appl. Surf. Sci.*, **386**, pp. 309–318.

49. Junyan, L., Li, W., Jingxian, B., and Ling, H. (2016). SiO_2-g-PS/fluoroalkylsilane composites for superhydrophobic and highly oleophobic coatings, *Colloids Surf. A: Physicochem. Eng. Aspects*, **507**, pp. 26–35.

50. Zhang, M., Zang, D., Shi, J., Gao, Z., Wang, C., and Li, J. (2015). Superhydrophobic cotton textile with robust composite film and flame retardancy, *RSC Adv.*, **5**, pp. 67780–67786.

51. Ramakrishna, S., Santhosh Kumar, K. S., Mathew, D., and Reghunadhan Nair, C. P. (2015). A robust, melting class bulk superhydrophobic material with heat-healing and self-cleaning properties, *Sci. Rep.*, **5**, pp. 18510 (1–11).

52. Sedai, B. R., Khatiwada, B. K., Mortazavian, H., and Blum, F. D. (2016). Development of superhydrophobicity in fluorosilane-treated diatomaceous earth polymer coatings, *Appl. Surf. Sci.*, **386**, pp. 178–186.

53. Zhang, L., Chen, S., and Dang, Z. M. (2016). Recycled polydopamine particles used for special wetting, *Mater. Chem. Phys.*, **181**, pp. 321–325.

54. Srinivasan, S., Praveen, V. K., Philip, R. T., and Ajayaghosh, A. (2008). Bioinspired superhydrophobic coatings of carbon nanotubes and linear π systems based on the "bottom-up" self-assembly approach, *Angew. Chem. Int. Ed.*, **47**, pp. 5750–5754.

55. Yang, C., Yang, X., Li, F., Li, T., and Cao, W. (2016). Controlled synthesis of hierarchical flower-like Sb_2WO_6 microspheres: photocatalytic and superhydrophobic property, *J. Ind. Eng. Chem.*, **39**, pp. 93–100.

56. Liu, Y., Cao, H., Chen, Y., Chen, S., and Wang, D. (2016). Self-assembled super-hydrophobic multilayer films with corrosion resistance on copper substrate, *RSC Adv.*, **6**, pp. 2379–2386.

57. Liao, K. S., Wan, A., Batteas, J. D., and Bergbreiter, D. E. (2008). Superhydrophobic surfaces formed using layer-by-layer self-assembly with aminated multiwall carbon nanotubes, *Langmuir*, **24**, pp. 4245–4253.

58. Gong, M., Xu, X., Yang, Z., Liu, Y., Lv, H., and Lv, L. (2009). A reticulate superhydrophobic self-assembly structure prepared by ZnO nanowires, *Nanotechnology*, **20**, pp. 165602 (1–7).

59. Du, R., Zhao, Q., Zheng, Z., Hu, W., and Zhang, J. (2016). 3D self-supporting porous magnetic assemblies for water remediation and beyond, *Adv. Energy Mater.*, **6**, pp. 1600473 (1–25).

60. Yu, L., Zhou, X., and Jiang, W. (2016). Low-cost and superhydrophobic magnetic foam as an absorbent for oil and organic solvent removal, *Ind. Eng. Chem. Res.*, **55**, pp. 9498–9506.

61. Prado, R., Beobide, G., Marcaide, A., Goikoetxea, J., and Aranzabe, A. (2010). Development of multifunctional sol–gel coatings: anti-reflection coatings with enhanced self-cleaning capacity, *Solar Energy Mater. Solar Cells*, **94**, pp. 1081–1088.

62. Zhang, X. X., Cai, S., You, D., Yan, L. H., Lv, H. B., Yuan, X. D., and Jiang, B. (2013). Template-free sol-gel preparation of superhydrophobic ORMOSIL films for double-wavelength broadband antireflective coatings, *Adv. Funct. Mater.*, **23**, pp. 4361–4365.

63. Yang, D. J., Li, J. P., Xu, Y., Wu, D., Sun, Y. H., Zhu, H. Y., and Deng, F. (2009). Direct formation of hydrophobic silica-based micro/mesoporous hybrids from polymethylhydrosiloxane and tetraethoxysilane, *Microporous Mesoporous Mater.*, **95**, pp. 180–186.

64. Nagappan, S., Lee, D. B., Seo, D. J., Park, S. S., and Ha, C. S. (2015). Superhydrophobic mesoporous material as a pH sensitive organic dye adsorbent, *J. Ind. Eng. Chem.*, **22**, pp. 288–295.

65. Nagappan, S., and Ha, C. S. (2014). Effect of sodium hydroxide on the fast synthesis of superhydrophobic powder from polymethylhydrosiloxane, *J. Coat. Sci. Technol.*, **1**, pp. 151–160.

66. Nagappan, S., Park, J. J., Park, S. S., and Ha, C. S. (2014). Preparation of superhydrophobic and transparent micro-nano hybrid coatings from polymethylhydroxysiloxane and silica ormosils, *Nano Convergence*, **1**, pp. 30 (1–10).

67. Nagappan, S., Jo, N. J., Lee, W. K., and Ha, C. S. (2017). Thermally stable superhydrophobic polymethylhydrosiloxane nanohybrids with liquid marble-like structure, *Macromol. Res.*, **25**, pp. 387–390.

68. Nagappan, S., and Ha, C. S. (2017). In-situ addition of graphene oxide for improving the thermal stability of superhydrophobic hybrid materials, *Polymer*, **116**, pp. 412–422.

69. Shirtcliffe, N. J., McHale, G., Newton, M. I., Perry, C. C., and Roach, P. (2005). Porous materials show superhydrophobic to superhydrophilic switching, *Chem. Commun.*, pp. 3135–3137.

70. Fu, Q., Wu, X., Kumar, D., Ho, J. W. C., Kanhere, P. D., Srikanth, N., Liu, E., Wilson, P., and Chen, Z. (2014). Development of sol-gel icephobic

coatings: effect of surface roughness and surface energy, *ACS Appl. Mater. Interfaces*, **6**, pp. 20685–20692.

71. Liu, S., Liu, X., Latthe, S. S., Gao, L., An, S., Yoon, S. S., Liu, B., and Xing, R. (2015). Self-cleaning transparent superhydrophobic coatings through simplesol–gel processing of fluoroalkylsilane, *Appl. Surf. Sci.*, **351**, pp. 897–903.

72. Rao, A. V., Latthe, S. S., Nadargi, D. Y., Hirashima, H., and Ganesan, V. (2009). Preparation of MTMS based transparent superhydrophobic silica films by sol–gel method, *J. Colloid Interface Sci.*, **332**, pp. 484–490.

73. Mahadik, S. A., Kavale, M. S., Mukherjee, S. K., and Rao, A. V. (2010). Transparent superhydrophobic silica coatings on glass by sol–gel method, *Appl. Surf. Sci.*, **257**, pp. 333–339.

74. Ahmad, N. A., Leo, C. P., and Ahmad, A. L. (2013). Synthesis of superhydrophobic alumina membrane: effects of sol–gel coating, steam impingement and water treatment, *Appl. Surf. Sci.*, **284**, pp. 556–564.

75. Wang, H., Ding, J., Xue, Y., Wang, X., and Lin, T. (2010). Superhydrophobic fabrics from hybrid silica sol-gel coatings: structural effect of precursors on wettability and washing durability, *J. Mater. Res.*, **25**, pp. 1336–1343.

76. Das, I., and De, G. (2015). Zirconia based superhydrophobic coatings on cotton fabrics exhibiting excellent durability for versatile use, *Sci. Rep.*, **5**, pp. 18503 (1–11).

77. Lei, Z., Deng, Y., and Wang, C. (2016). Ambient-temperature fabrication of melamine based sponges coated with hydrophobic lignin shells by surface dip adsorbing for oil/water separation, *RSC Adv.*, **6**, pp. 106928–106934.

78. Liu, Y., Zhang, K., Yao, W., Liu, J., Han, Z., and Ren, L. (2016). Bioinspired structured superhydrophobic and superoleophilic stainless steel mesh for efficient oil-water separation, *Colloids Surf. A: Physicochem. Eng. Aspects*, **500**, pp. 54–63.

79. Zhu, J., Liu, B., Li, L., Zeng, Z., Zhao, W., Wang, G., and Guan, X. (2016). Simple and green fabrication of a superhydrophobic surface by one-step immersion for continuous oil/water separation, *J. Phys. Chem. A*, **120**, pp. 5617–5623.

80. Yan, W., Liu, H., Chen, T., Sun, Q., and Zhu, W. (2016). Fast and low-cost method to fabricate large-area superhydrophobic surface on steel

substrate with anticorrosion and anti-icing properties, *J. Vac. Sci. Technol. A*, **34**, p. 041401.

81. Si, Y., and Guo, Z. (2016). Eco-friendly functionalized superhydrophobic recycled paper with enhanced flame-retardancy, *J. Colloid Interface Sci.*, **477**, pp. 74–82.

82. Gao, Z., Zhai, X., Liu, F., Zhang, M., Zang, D., and Wang, C. (2015). Fabrication of TiO_2/EP super-hydrophobic thin film on filter paper surface, *Carbohydr. Polym.*, **128**, pp. 24–31.

83. Chen, S., Li, X., Li, Y., and Sun, J. (2015). Intumescent flame-retardant and self-healing superhydrophobic coatings on cotton fabric, *ACS Nano*, **9**, pp. 4070–4076.

84. Cao, C., Ge, M., Huang, J., Li, S., Deng, S., Zhang, S., Chen, Z., Zhang, K., Al-Deyab, S. S., and Lai, Y. (2016). Robust fluorine-free superhydrophobic PDMS–ormosil@fabrics for highly effective self-cleaning and efficient oil–water separation, *J. Mater. Chem. A*, **4**, pp. 12179–12187.

85. Li, X., Yan, P., Li, H., and Gao, X. (2016). Fabrication of tunable, stable, and predictable superhydrophobic coatings on foam ceramic materials, *Ind. Eng. Chem. Res.*, **55**, pp. 10095–10103.

86. Liu, W., Xu, Q., Han, J., Chen, X., and Min, Y. (2016). A novel combination approach for the preparation of superhydrophobic surface on copper and the consequent corrosion resistance, *Corros. Sci.*, **110**, pp. 105–113.

87. Wang, H., Zhu, Y., Hu, Z., Zhang, X., Wu, S., Wang, R., and Zhu, Y. (2016). A novel electrodeposition route for fabrication of the superhydrophobic surface with unique self-cleaning, mechanical abrasion and corrosion resistance properties, *Chem. Eng. J.*, **303**, pp. 37–47.

88. Li, H., Yu, S., Han, X. X., and Zhao, Y. (2016). A stable hierarchical superhydrophobic coating on pipeline steel surface with self-cleaning, anticorrosion, and anti-scaling properties, *Colloids Surf. A: Physicochem. Eng. Aspects*, **503**, pp. 43–52.

89. Lin, Y., Shen, Y., Liu, A., Zhu, Y., Liu, S., and Jiang, H. (2016). Bio-inspiredly fabricating the hierarchical 3D porous structure superhydrophobic surfaces for corrosion prevention, *Mater. Des.*, **103**, pp. 300–307.

90. Xu, N., Sarkar, D. K., Chen, X., and Tong, W. P. (2016). Corrosion performance of superhydrophobic nickel stearate/nickel hydroxide thin films on aluminum alloy by a simple one-step electrodeposition process, *Surf. Coat. Technol.*, **302**, pp. 173–184.

91. Mahadik, S. A., Mahadik, D. B., Kavale, M. S., Parale, V. G., Wagh, P. B., Barshilia, H. C., Gupta, S. C., Hegde, N. D., and Rao, A. V. (2012).

Thermally stable and transparent superhydrophobic sol–gel coatings by spray method, *J. Sol-Gel Sci. Technol.*, **63**, pp. 580–586.

92. Cirisano, F., Benedetti, A., Liggieri, L., Ravera, F., Santini, E., and Ferrari, M. (2016). Amphiphobic coatings for antifouling in marine environment, *Colloids Surf. A: Physicochem. Eng. Aspects*, **505**, pp. 158–164.

93. Ramakrishna, S., Santhosh Kumar, K. S., Mathew, D., and Nair, C. P. R. (2015). Long-living, stress- and pH-tolerant superhydrophobic silica particles via fast and efficient urethane chemistry; facile preparation of self-recoverable SH coatings, *J. Mater. Chem. A*, **3**, pp. 1465–1475.

94. Alexander, S., Eastoe, J., Lord, A. M., Guittard, F., and Barron, A. R. (2016). Branched hydrocarbon low surface energy materials for superhydrophobic nanoparticle derived surfaces, *ACS Appl. Mater. Interfaces*, **8**, pp. 660–666.

95. Wang, H., Wang, C., Gao, D., Li, M., Zhu, Y., and Zhu, Y. (2016). Heat/durability resistance of the superhydrophobic PPS-based coatings prepared by spraying non-fluorinated polymer solution, *Colloid Polym. Sci.*, **294**, pp. 1519–1527.

96. Li, J., Wu, R., Jing, Z., Yan, L., Zha, F., and Lei, Z. (2015). One-step spray-coating process for the fabrication of colorful superhydrophobic coatings with excellent corrosion resistance, *Langmuir*, **31**, pp. 10702–10707.

97. Ogihara, H., Xie, J., Okagaki, J., and Saji, T. (2012). Simple method for preparing superhydrophobic paper: pray-deposited hydrophobic silica nanoparticle coatings exhibit high water-repellency and transparency, *Langmuir*, **28**, pp. 4605–4608.

98. Xu, L., Karunakaran, R. G., Guo, J., and Yang, S. (2012). Transparent, superhydrophobic surfaces from one-step spin coating of hydrophobic nanoparticles, *ACS Appl. Mater. Interfaces*, **4**, pp. 1118–1125.

99. Huang, D. J., and Leu, T. S. (2013). Fabrication of high wettability gradient on copper substrate, *Appl. Surf. Sci.*, **280**, pp. 25–32.

100. Söz, C. K., Yilgör, E., and Yilgör, I. (2016). Simple processes for the preparation of superhydrophobic polymer surfaces, *Polymer*, **99**, pp. 580–593.

101. Nagappan, S., Park, J. J., Park, S. S., Lee, W. K., and Ha, C. S. (2013). Bio-inspired, multi-purpose and instant superhydrophobic lotus leaf powder hybrid micro-nanocomposites for selective oil spill capture, *J. Mater. Chem., A*, **1**, pp. 6761–6769.

102. Nagappan, S., Park, J. H., Sung, A. R., and Ha, C. S. (2014). Superhydrophobic hybrid micro-nanocomposites for non-stick and self-cleaning coatings, *Compos. Interfaces*, **21**, pp. 597–609.

103. Nagappan, S., Park, S. S., and Ha, C. S. (2014). Superhydrophobic and self-cleaning natural leaf powder/poly(methylhydroxysiloxane) hybrid micro-nanocomposites, *Macromol. Res.*, **22**, pp. 843–852.

104. Nagappan, S., Sung, A. R., and Ha, C. S. (2014). Effect of temperature on the lotus leaf powder and the preparation of bio-based silica hybrid micro-nanocomposites, *J. Bio-based Mater. Bio-energy*, **8**, pp. 175–183.

105. Nagappan, S., Park, S. S., Tapaswi, P. K., Rao, K. M., Ha, C. S., and Hwang, T. S. (2015). *Camellia japonica*-polysiloxane based superhydrophobic hybrid powder for the selective adsorption of metal ions from a mixture of metal ions in artificial sea water, *J. Porous Mater.*, **22**, pp. 229–238.

106. Nagappan, S., and Ha, C. S. (2015). Superhydrophobic hybrid micro-nanocomposites with various applications, *Macromol. Symp.*, **358**, pp. 202–211.

107. Esteves, A. C. C., Luo, Y., van de Put, M. W. P., Carcouët, C. C. M., and de With, G. (2014). Self-replenishing dual structured supehydrophobic coatings prepared by drop-casting of an all-in-one dispersion, *Adv. Fun. Mater.*, **24**, pp. 986–992.

108. Wang, H., Sun, F., Wang, C., Zhu, Y., and Wang, H. (2016). A simple drop-casting approach to fabricate the super-hydrophobic PMMA-PSF-CNFs composite coating with heat-, wear- and corrosion-resistant properties, *Colloid Polym. Sci.*, **294**, pp. 303–309.

109. Park. E. J., Kim, K. D., Yoon, Y. S., Jeong, M. G., Kim, D. H., Lim, D. C., Kim, Y. H., and Kim, Y. D. (2014). Fabrication of conductive, transparent and superhydrophobic thin films consisting of multi-walled carbon nanotubes, *RSC Adv.*, **4**, pp. 30368–30374.

110. Sas, I., Gorga, R. E., Joines, J. A., and Thoney, K. A. (2012). Literature review on superhydrophobic self-cleaning surfaces produced by electrospinning, *J. Polym. Sci. Part B: Polym. Phys.*, **50**, pp. 824–845.

111. Li, X., Yu, X., Cheng, C., Deng, L., Wang, M., and Wang, X. (2015). Electrospun superhydrophobic organic/inorganic composite nanofibrous membranes for membrane distillation, *ACS Appl. Mater. Interfaces*, **7**, pp. 21919–21930.

112. Shami, Z., Amininasab, S. M., and Shakeri, P. (2016). Structure–property relationships of nanosheeted 3D hierarchical roughness MgAl–layered double hydroxide branched to an electrospun porous nanomembrane:

a superior oil-removing nanofabric, *ACS Appl. Mater. Interfaces*, **8**, pp. 28964–28973.

113. Arslan, O., Aytac, Z., and Uyar, T. (2016). Superhydrophobic, hybrid, Electrospun cellulose acetate nanofibrous mats for oil/water separation by tailored surface modification, *ACS Appl. Mater. Interfaces*, **8**, pp. 19747–19754.

114. Dong, F., Zhang, M., Huang, W., Zhou, L., Wong, M.S., and Wang, Y. (2015). Superhydrophobic/hydrophobic nanofibrous network with tunable cell adhesion: fabrication, characterization and cellular activities, *Colloids Surf. A: Physicochem. Eng. Aspects*, **482**, pp. 718–723.

115. Pan, S., Kota, A. K., Mabry, J. M., and Tuteja, A. (2013). Superomniphobic surfaces for effective chemical shielding, *J. Am. Chem. Soc.*, **135**, pp. 578–581.

116. Kumar, V., and Sharma, N. N. (2015). Synthesis of hydrophilic to superhydrophobic SU8 surfaces, *J. Appl. Polym. Sci.*, **132**, pp. 41934 (1–10).

117. Liu, Y., Liu, Y., Hu, H., Liu, Z., Pei, X., Yu, B., Yan, P., and Zhou, F. (2015). Mechanically induced self-healing superhydrophobicity, *J. Phys. Chem. C*, **119**, pp. 7109–7114.

118. Li, L., Breedveld, V., and Hess, D. W. (2012). Creation of superhydrophobic stainless steel surfaces by acid treatments and hydrophobic film deposition, *ACS Appl. Mater. Interfaces*, **4**, pp. 4549–4556.

119. Gao, J., Li, Y., Li, Y., Liu, H., and Yang, W. (2012). Fabrication of superhydrophobic surface of stearic acid grafted zinc by using an aqueous plasma etching technique, *Cent. Eur. J. Chem.*, **10**, pp. 1766–1772.

Chapter 5

Applications of Superhydrophobic Organic-Inorganic Nanohybrids

5.1 Introduction

Organic-inorganic hybrid surfaces were prepared by simple surface treatment with one or multisteps of modifications using different hydrophobic materials to exhibit excellent superhydrophobicity on various substrates. Moreover, the fabricated substrates have attracted considerable attention for various applications, such as oil sorption and separation from water, anticorrosion, anti-icing, photocatalysis, and textiles. The superhydrophobic coatings are commercially important in a range of industrial products because of the increase in the lifetime of the products for prolonged and repeated use.

5.2 Applications of Superhydrophobic Organic-Inorganic Nanohybrids

5.2.1 Oil Sorption and Separation

Superhydrophobic surfaces are used for the sorption and separation of large quantities of oils from a water surface. This

Hydrophobic and Superhydrophobic Organic-Inorganic Nanohybrids
Chang-Sik Ha and Saravanan Nagappan

ISBN 978-981-4774-68-0 (Hardcover), 978-1-351-20607-5 (eBook)
www.panstanford.com

is due to the excellent water-repellent and oil-like properties of superhydrophobic substrates, which absorb or separate oil selectively from the water surface. Moreover the superhydrophobic substrates also show excellent recyclability and high separation efficiency of the substrates. Various substrates, such as sponge, foam, cotton, textile, and paper, were rendered superhydrophobic using simple steps and used for the sorption and separation of oils from a water surface [1–7]. This is due to the large amounts of sorption or separation of oil from the water surface and the easy recycling of the oils and reusability of the substrates. Recently, magnetic-based superhydrophobic surfaces have attracted considerable attention in oil sorption or separation from water surface due to the magnetic-responsive behavior of the materials or surfaces. Nagappan and Ha reviewed the various methods of synthesis, fabrication, and applications of superhydrophobic surfaces based on magnetic materials [8].

The oil absorption capacity, k, was calculated in two ways, by either the mass or the volume ratio, based on the following equations:

$$k = (W_s - W_i)/W_i, \quad (5.1)$$

$$k = (W_s - W_i)/\rho_o V_s, \quad (5.2)$$

where W_i and V_s are the weight and volume of the superhydrophobic substrate before absorption, respectively; W_s is the total weight of the superhydrophobic substrate after the absorption of oil and organic solvent at the saturation; and ρ_o is the density of oil and organic solvent absorbed.

The separation efficiency (R) was calculated using the following equation:

$$R\,(\%) = (1 - C_p/C_o) \times 100, \quad (5.3)$$

where C_o and C_p are the oil concentrations before and after separation, respectively.

The organic-inorganic hybrid suspension prepared from lotus leaf powder (LLP) and functional siloxane-based materials was used to fabricate the superhydrophobic melamine sponge and for the oil spill capture applications [9].

The melamine sponge was precleaned in acetone, ethanol, and water and dried prior to use. The cleaned melamine sponge was immersed in the hybrid suspension, followed by surface treatment with polydimethylsiloxane (PDMS) and drying at room temperature, resulting in excellent superhydrophobicity (Fig. 5.1) [9]. The sponge showed a silvery mirror image by dipping into water, which reflects the excellent water repellence of the sponge (Fig. 5.1). Furthermore, the superhydrophobic sponge was used for the sorption of various oils and organic solvents from the water surface as well as below the water. The superhydrophobic melamine sponge exhibited the excellent sorption of chloroform and various oils and organic solvents and was also used for continuous sorption for several cycles (Fig. 5.1) [9].

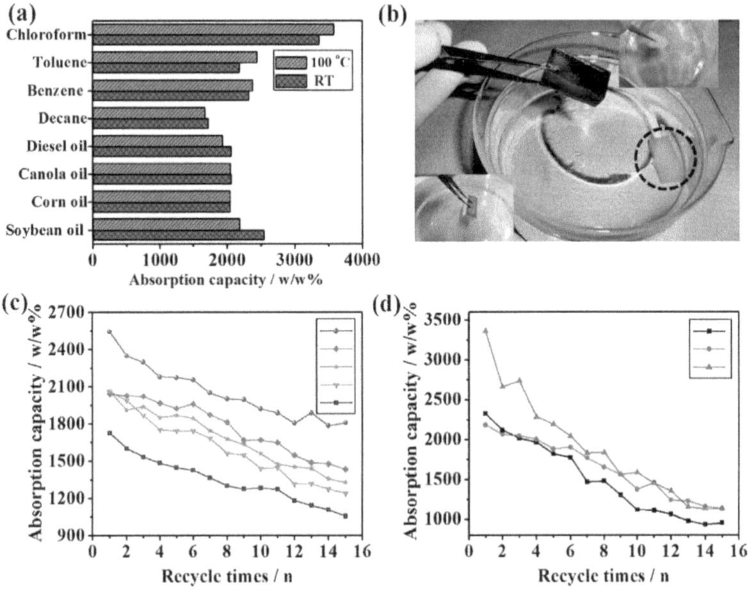

Figure 5.1 (a) Absorption of various oils and organic solvents by the superhydrophobic hybrid/PDMS sponge cured at room temperature (RT) and 100°C. (b) Optical image of the hybrid/PDMS sponge after the absorption of diesel oil. The complete superhydrophilic properties of the pristine sponge are illustrated at the top right of the inset and at the bottom (black dashed circle) of the water bath. The bottom left of the inset shows the complete superhydrophobic properties of the hybrid/PDMS sponge when immersed in water (silvery shine). (c, d) Graphs of the absorption capacity of various oils (soybean •, corn ◆, canola ✱, diesel ▼, and decane ■) and various solvents (benzene ■, toluene •, and chloroform ▲) of the hybrid-loaded and surface-treated sponge cured at RT. Reprinted from Ref. [9] with permission from The Royal Society of Chemistry.

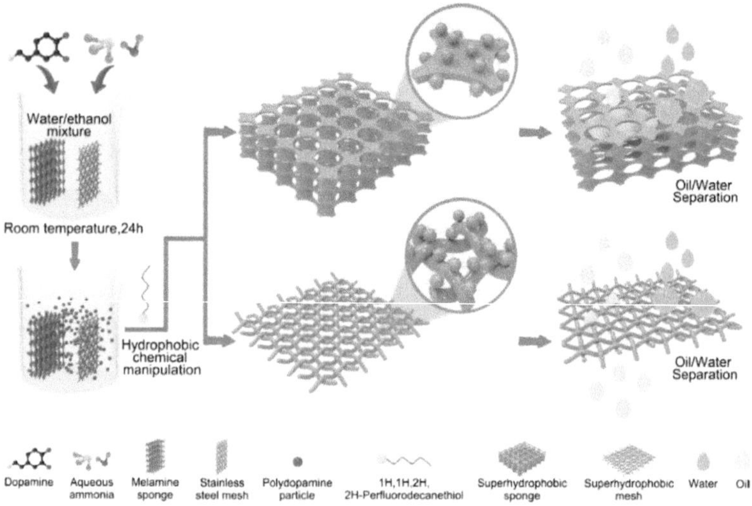

Scheme 5.1 Schematic illustration of superhydrophobic modification of commercial MF sponge and stainless steel mesh as examples for potential use in oil–water separation. Reprinted from Ref. [10]. Copyright (2016), with permission from Elsevier.

A series of superhydrophobic surfaces from a melamine-formaldehyde (MF) sponge, stainless steel mesh (300 mesh size), nylon netting, and cotton cloth substrates were developed by modifying the substrate surface with polydopamine (PDA) and 1H,1H,2H,2H-perfluorodecanethiol (Scheme 5.1) [10]. The substrates exhibited excellent superhydrophobic stability because of the highly hydrophobic materials coated on the 3D porous structure. The porous superhydrophobic substrates also showed excellent sorption and separation efficiency of various oils and organic solvents from the water surface.

A superoleophobic hybrid aerogel was prepared by freeze-drying sodium alginate and nanofibrillated cellulose (NFC) using Ca^{2+} ions as the crosslinking agent [11]. First, the sodium alginate (2 wt%) solution and NFC (1 wt%) suspension were prepared separately and mixed together by continuous stirring until the formation of a homogeneous dispersion and the mixture was frozen in a Petri dish. The resulting aerogel was induced to form more crosslinked networks by immersing in a calcium chloride (4 wt%) solution for 10 h followed by washing off the excess calcium ions and drying the aerogel. The prepared sodium alginate/NFC aerogel showed a highly 3D porous structure with hydrophilic and underwater superoleophobic properties. Moreover, the excellent underwater superoleophobic and hydrophilic properties of the aerogel may lead to use of the aerogel for the separation of oils from water as well as from real seawater. The authors demonstrated excellent stability of the aerogel over 30 days immersed in seawater as well as the high separation efficiency (up to 99.65%) of various oils from water and good reusability (at least 40 cycles).

The core–shell type of well-ordered and uniform-size magnetic organic-inorganic hybrid nanoparticles were fabricated by first synthesizing the Fe_3O_4 magnetic nanoparticles, followed by surface treatment with PDA and 3-mercaptopropyl trimethoxysilane and loading with silver nanoparticles [12]. The synthesized magnetic hybrid nanoparticles exhibited excellent superhydrophobicity and formed a liquid marble by rolling a water droplet on the highly hydrophobic nanoparticle bed. The electrostatic attraction of the functional nanoparticles and water droplet can form a thin of powder film of the water droplet surface and show a liquid marble droplet. The magnetic liquid marble showed the sorption of oils and organic solvents at the water surface. Moreover, the functional hybrid magnetic nanoparticles were also actuated easily by a magnetic bar due to the excellent magnetic properties of the material. The results highlight the excellent magnetic-responsive behavior of the hybrid materials that is suitable for oil spill capture application.

Sukamanchi et al. synthesized (leafhopper-type) silica nanoparticles that were modified by the catalytic grafting of stearic acid, and a superhydrophobic surface was fabricated on a glass substrate using the drop-casting method [13]. The fabricated surface showed excellent superhydrophobic (water contact angle [WCA] = 165°–172°) stability against various pH values (acidic and basic). The superhydrophobic silica nanoparticles were also superoleophilic and showed oil spill capture properties. A magnetic superhydrophobic foam was fabricated using a low-cost polyethylene (PE) shock absorption foam [14]. The magnetic superhydrophobic foam exhibited excellent absorption of various oils and organic solvents from the water surface on account of the superoleophilic property of the magnetic foam. The sorbed oils and organic solvents were collected easily by simple squeezing the magnetic superhydrophobic foam and used for the continuous capture of oils and organic solvents. The surface could be recycled for several cycles due to the magnetic-responsive behavior of the superhydrophobic foam.

Zhu et al. fabricated a superhydrophobic stainless steel mesh by treating the steel mesh surface by the one-step immersion in a ferric stearate solution followed by curing the substrate at room temperature [7]. The substrate exhibited excellent superhydrophobicity on the stainless steel substrate and was used for the absorption of various oils from a water surface. The superhydrophobic stainless steel substrate was also used for the continuous separation of oils using a simple laboratory setup with a pressure controller (Fig. 5.2).

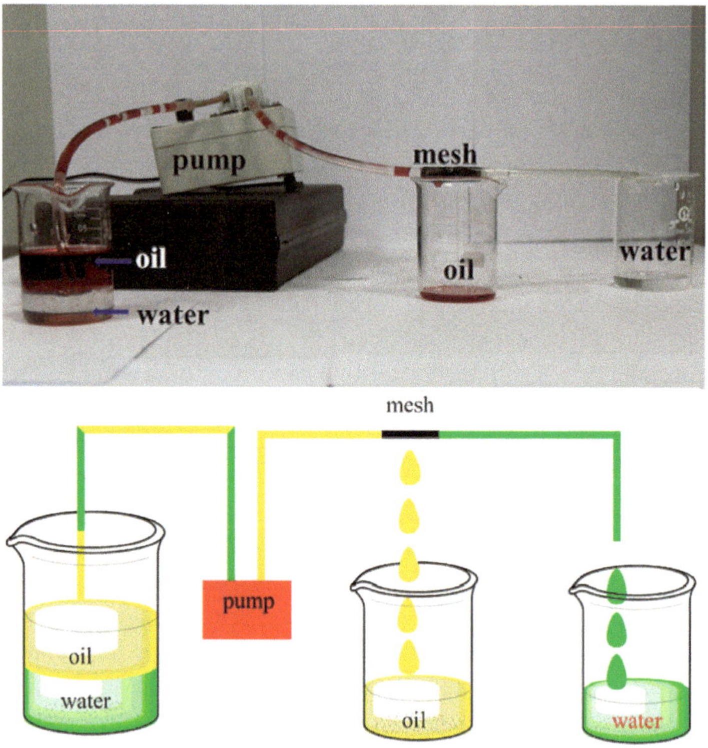

Figure 5.2 Photograph of the experimental apparatus continuously collecting pure oil from the oil–water mixture and schematic diagram of the oil–water separation apparatus. Reprinted from Ref. [7]. Copyright (2016), with permission from the American Chemical Society.

5.2.2 Anticorrosion

The superhydrophobic coating substrate showed better corrosion resistance than the hydrophobic coating substrate because of the excellent water-repellent property of the coating materials. Several superhydrophobic coating materials and substrates were developed and used for anticorrosion applications.

In general, the anticorrosion behavior of a substrate is checked by immersing the coated substrate in a corrosive developing solution (3.5% sodium chloride [NaCl] solution) for various dipping times followed by mild washing with deionized water and curing. The corrosion response of the substrate is measured on an electrochemical workstation. The corrosion behavior was measured in a three-electrode system, where the superhydrophobic coated substrate, platinum, and Ag/AgCl are used as working, auxiliary, and reference electrodes (Fig. 5.3) [15].

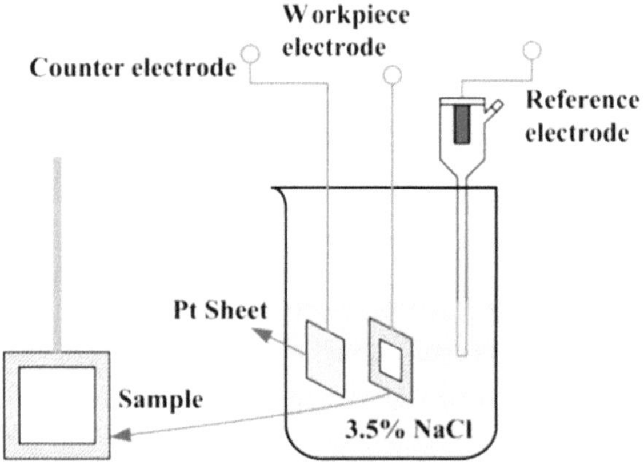

Figure 5.3 Schematic diagram of the three-electrode electrochemical cell. Reprinted from Ref. [15]. Copyright (2016), with permission from Elsevier.

The corrosion inhibition efficiency (η_p) is calculated from the following equation:

$$(\eta_p) = [(i^0_{corr} - i_{corr})/i^0_{corr}) \times 100, \tag{5.4}$$

where i^0_{corr} and i_{corr} are the corrosion current densities without and with the inhibitor (film), respectively. The higher corrosion inhibition efficiency would show better anticorrosion behavior of the substrate.

A hierarchical micro-/nanoscale structure on a Ti_6Al_4V titanium alloy substrate was obtained by sandblasting and two-step chemical etching followed by surface treatment with fluoroalkylsilane (FAS), which rendered the substrate superhydrophobic [15]. The superhydrophobicity on the aluminum substrate depended on the concentration of sodium hydroxide (NaOH) solution used for the hydrolysis and condensation of FAS. The surface contact angle (CA) increased with increasing NaOH solution concentration due to enhanced hierarchical surface morphology and the formation of low-surface-energy materials (LSEMs) (FAS) on the substrate. The FAS-modified substrate showed a higher open-circuit potential (OCP) of 0.27 V than the unmodified substrate OCP (0.02 V due to the nonconductive air cushion trapped on the interface) and good corrosion behavior of the substrate.

The FAS-treated substrate also showed a 3 orders of magnitude lower corrosion current density (i_{corr}) than the unmodified substrate. Similar results were also obtained by electrochemical impedance spectroscopy (EIS) for the superhydrophobic substrate with a 3.5 orders of magnitude higher impedance modulus than the unmodified substrate. The increased corrosion protection is due to the prevention of Cl^- migration by the superhydrophobic substrate. The results suggest the excellent corrosion resistance behavior of the superhydrophobic substrate.

The superhydrophobic steel substrate was fabricated by first preparing hydrophobically modified silica nanoparticles using 1H,1H,2H,2H-perfluoroalkyltriethoxysilane (97%) (FAS-*X*) and mixed with polymethylmethacrylate (PMMA) acrylic powder [16]. The coating solution was prepared by mixing PMMA, FAS-*X*-treated silica nanoparticles, and tetrahydrofuran (THF) with a mass ratio of 1:x:90 for 3 h at room temperature and spray-coated on a steel substrate. The coated substrate exhibited excellent superhydrophobicity and anticorrosion properties.

A superhydrophobic organic-inorganic hybrid aluminum substrate was fabricated for anticorrosion applications by the electrochemical deposition of a mixture of an ethanoic solution of different ratios of stearic acid and nickel nitrate hexahydrate solution with an applied direct current (DC) power supply of 20 V for 10 min at a cathode to anode distance of 15 cm (Fig. 5.4) [17]. The fabricated substrate was hydrophobic and superhydrophobic depending on the concentration of the Ni^{2+}/sodium alginate solution. Micronanohierarchical nickel stearate particles were deposited on the aluminum surface by increasing the concentration of the Ni^{2+}/sodium alginate solution, which also increases the surface CA and surface roughness. Nickel ions are used as a corrosion inhibitor, whereas the long alkyl stearate molecule can prevent water contact on the coated substrate. The combination of both the advantages of corrosion inhibition and excellent water-repellent properties of the fabricated aluminum substrate resulted in excellent anticorrosion behavior. The corrosion resistance of the substrate was examined using an automatic calculated machine, which analyzed the corrosion responsive behavior of the materials and coated substrates.

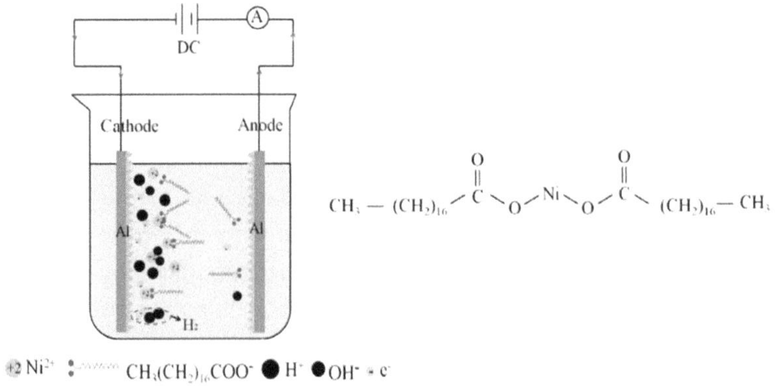

Figure 5.4 (a) Schematic presentation of the electrodeposition process and (b) molecular structure of nickel stearate. Reprinted from Ref. [17]. Copyright (2016), with permission from Elsevier.

The polarization curve of the chemically cleaned aluminum substrate (CA of 54° ± 2°) showed a corrosion current density of 16.06 ± 0.19 μA/cm^2, which was reduced dramatically to 0.78 ± 0.08 μA/cm^2 by the deposition of a lower molar ratio of Ni^{2+}/sodium alginate (0.015 M, CA = 135° ± 1°) solution. The increase in the molar ratio of the Ni^{2+}/sodium alginate (0.4 M, CA = 160° ± 1°) solution for deposition on an aluminum substrate leads to higher superhydrophobicity, which further reduces the corrosion current density to 0.01 ± 0.006 μA/cm^2. These results suggest that the superhydrophobic substrate with nickel stearate deposition showed better anticorrosion resistance than the hydrophobic and chemically treated hydrophilic aluminum substrates. The polarization resistance of the chemically cleaned aluminum substrate and the Ni^{2+}/sodium alginate solution–modified hydrophobic and superhydrophobic substrates were 2.88 ± 0.62 kΩ·cm^2, 59.29 ± 19.69 kΩ·cm^2, and 816.9 ± 203.96 kΩ·cm^2, respectively. The polarization resistance is inversely proposal to the current density. Decreasing the current density would increase the polarization resistance of the substrates. The increased polarization resistance and decreased current density was attributed to the shorter distance between the hierarchical micronanostructures, and the hydrophobicity of the substrate would reduce the affinity of water droplets to the surface.

Liu et al. fabricated a superhydrophobic copper substrate by first modifying the copper foil surface by abrading with different grades of emery paper followed by etching in an ammonia solution (50 mL) at room temperature for 20 h, washing with alcohol and deionized water and calcining in air at 340°C for 10 min. The copper substrate was used as an electrode material [18], which was modified using a 0.1 mol/L ethanol solution of stearic acid at room temperature for 3 h. The surface property of the modified copper foil was checked at various etching times followed by surface modification. The bare copper substrate was hydrophilic (76.5°), whereas the surface became hydrophobic to superhydrophobic (CA = 132.7°, 148.2°, 150.3°, 157.6°, and 152.4°) by increasing the etching time of the substrate from 5 h to 25 h.

A saturated superhydrophobic CA was obtained on the copper foil surface at an etching time of 20 h. The fabricated substrate exhibited excellent corrosion resistance, which was confirmed by immersing the surface-modified copper foil substrate in a 3.5% NaCl solution at room temperature when the corrosion behavior was studied on an electrochemical workstation. The superhydrophobic copper film showed three times higher impedance than that of the pristine copper substrate. The impedance spectra showed a linear relationship with a slope close to –1 in the higher-frequency range, indicating the excellent anticorrosion behavior of the superhydrophobic copper substrate. Superhydrophobic anticorrosion surfaces were also prepared on various substrates using different materials [19–23].

5.2.3 Anti-icing

Icing is a major problem for refrigerator manufacturers, wind power mills, automobiles, aerospace industries, solar roof panels, and highways. Ice formation on these surfaces reduces the performance of the products and stored materials. In general, a hydrophilic surface can show great affinity to adhere to frost followed by the formation of stable ice crystals on the substrate due to the attraction of the frost crystals and continuous deposition of a frost crystal layer, which is solidified on the surface. The deposited ice should be removed promptly to improve the product performance and reduce the electricity consumption and other properties. Anti-icing is an important property for preventing the ice and frost formation on a substrate. Various chemicals were developed and used for de-icing the solid ice crystals from a substrate. On the other hand, the prevention of icing on a solid substrate is a challenging approach due to the difficulty in controlling the deposition of frost on a substrate and removing it from the substrate.

Two effective strategies are used to control ice adhesion, passive strategies (e.g., surface coating) and active strategies (e.g., infrared hangars, weeping wings, and electrical heating elements). Passive strategies are preferred due to the simple surface modification of the coating surfaces and the ease of handling the coating materials. On the other hand, active strategies are quite energy consuming, cost effective, and labor intensive. A superhydrophobic coating was introduced for anti-icing applications on various substrates. The results suggest that the formation of superhydrophobicity on a substrate results in better anti-icing performance than on hydrophobic and hydrophilic substrates. The water-repellent and self-cleaning properties of a superhydrophobic surface can resist frost formation and increase the performance of the coated products. A superhydrophobic anti-icing coating and substrates were developed through simple changes in the surface properties and chemical nature of the materials.

Fu et al. prepared superhydrophobic substrates by a sol-gel method using methyltriethoxysilane (MTES), 3-glycidyloxypropyl trimethoxysilane (GLYMO), and 1H,1H,2H,2H-perfluorodecyltriethoxysilane (FAS-17b) for anti-icing coatings [24]. The ice adhesion strength depended on the surface energy of the substrates. Two substrates with similar surface roughness were prepared. One substrate with a lower surface energy showed less ice adhesion than the other substrates with higher surface energies. The roughness of the fabricated substrate is also important for ice adhesion. The surface with higher roughness without a hydrophobic coating exhibits stronger wetting phenomena and a water drop on the surface freezes easily due to the mechanical interlocking effect of the substrate and ice, which induces stronger bonding to the substrate. The rougher the surface, the greater the anchoring effect. The hydrophobic FAS-17b surface treatment on rougher substrates showed less ice adhesion than on the untreated samples, even at subzero temperature. The authors also suggested that the surface superhydrophobicity and water repellence at room temperature are not the same by comparing with the subzero temperature. On the other hand, the FAS-17b-treated substrate exhibited excellent water repellence and ice adhesion properties even at subzero temperatures (−10°C). These results highlight the excellent anti-icing properties of the prepared superhydrophobic surface using a FAS-17b treatment due to the lower surface energy of the FAS-17b.

Beemer et al. fabricated durable organic-inorganic hybrid PDMS gels by a hydrosilylation reaction of vinyl-terminated PDMS (v-PDMS) and hydride-terminated PDMS (h-PDMS) with the addition of a nonreactive trimethyl-terminated PDMS (t-PDMS) plasticizer [25]. The prepared polysiloxane gel showed a low shear modulus and weakly crosslinked PDMS gels; the fabricated glass substrates also showed durable and ultralow ice adhesion strength.

An organic-inorganic hybrid superhydrophobic steel substrate was fabricated by spray-coating a mixture of PMMA and hydrophobically modified silica nanoparticles [26]. Hydrophobic silica nanoparticles were obtained by mixing the silica precursor (tetraethoxysilane [TEOS]) with an ammonia hydroxide solution and ethanol followed by the addition of hydrophobic fluorosilane. The superhydrophobic substrate was fabricated by mixing different weights of PMMA and hydrophobic silica nanoparticles and spray coating them on a steel substrate. The fabricated steel substrate exhibited excellent anti-icing behavior due to excellent water repellence of the substratee.

The anti-icing properties of the fabricated substrate were assessed using a cold-water dripping test and a condensing test. The cold water dripping test was carried out in an anti-icing chamber at an inner temperature of −20°C. The steel substrate was placed in the chamber and a water droplet (size of ~50 μL, at 0°C) was dropped on the substrate using a syringe (controlled the temperature of water). The freezing and bouncing properties of the water droplet were monitored on the superhydrophobic substrate. The fabricated substrate showed excellent water repellence; the water droplets moved away from the substrate easily. A few distinct frozen spots were observed on the steel substrate using this approach. On the other hand, a condensation test was carried out in a humidity chamber by placing the superhydrophobic steel substrate and reducing the temperature from 40°C to −20°C. The anti-icing behavior on the superhydrophobic steel substrate was observed from the condensation behavior of a liquid droplet in the chamber.

A thin moisture film was observed on the superhydrophobic steel substrate, which indicates a delay of icing on the substrate, whereas a foggy film covered the untreated steel substrate, suggesting the complete icing of the substrate (Fig. 5.5). These results suggest that the superhydrophobic substrate has much better anti-icing properties with delayed ice formation on the substrate due to overcooling of the substrate than the untreated steel substrate.

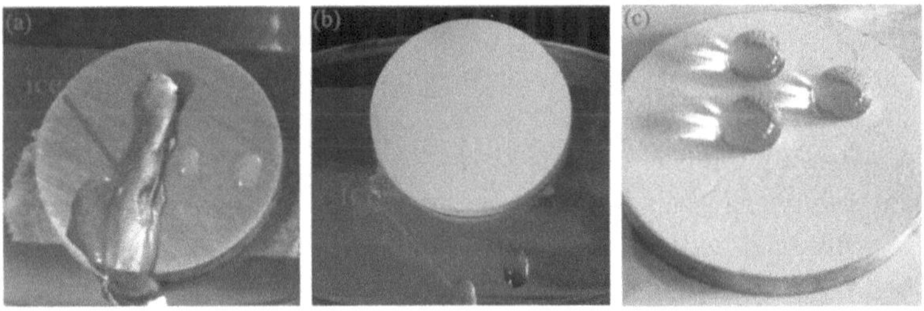

Figure 5.5 Water dripping on (a) steel substrate and (b) PMMA/fluorinated silica nanoparticle–modified steel substrate. (c) Water droplets on the surface of PMMA/fluorinated silica nanoparticle–modified steel substrate after a cool-water dripping test. Reprinted from Ref. [26]. Copyright (2016), with permission from Elsevier.

A flexible superhydrophobic substrate was also fabricated by patterning a lotus leaf structure using PDMS and surface treatment with zinc oxide nanoparticles and hydrophobic heptadecafluorodecyl tripropoxy silane (FAS-17d) [27]. The fabricated substrate exhibited robust superhydrophobicity, excellent flexibility, and anti-icing properties. The superhydrophobicity and anti-icing properties were maintained on the fabricated substrate, even after several bending and icing-melting cycles. Several other studies reported the superior anti-icing property of the superhydrophobic surfaces to hydrophobic and hydrophilic surfaces. The excellent water repellence of the superhydrophobic surface would delay ice formation on the fabricated substrate due to the repellence of the substrate surface for polar groups [28–30]. The substrate surface with superhydrophobic and slippery behavior would show better anti-icing behavior than superhydrophobic surfaces. This is due to the slippery nature of the fabricated superhydrophobic surface that could easily repel and remove the deposited fog and tiny water droplets from the substrate [31–33].

5.2.4 Antifouling Coatings

An antifouling coating is an important surface coating for preventing the fouling of biological entities on a substrate surface [34–36]. This coating is applied specifically to the outer surface of the hull of a ship or boat and improves the fouling resistance and enhances the life of the product [37–39]. The fouling of various kinds of isolated proteins and biopolymers occur continuously on a substrate that weaken the surface properties of the materials (Table 5.1 and Fig. 5.6) [34–38].

Several studies were carried out for the release and prevention of fouling of biological entities on a substrate [39–44]. Generally, low-surface-energy fluorine-based materials are used widely to prevent the fouling of biological entities due to the excellent water-repellent behavior of fluorine compounds [45–48]. Similarly, various other kinds of polymers and organic-inorganic nanohybrid materials with low surface energy were also used for the antifouling coatings [49, 50].

Recently, several studies have examined the antifouling performance of superhydrophobic surfaces [51–54]. The superhydrophobic surfaces exhibited excellent antifouling performance due to their water repellence, nonstick, and self-cleaning properties, which prevent the adhesion of biological molecules or proteins on the surface [55, 56].

Table 5.1 Characteristics of main marine macroorganism species

Groups	Algae (plants)	Invertebrates (animals)							
Subgroups	(a) green, (b) brown and (c) red	Hard shell organisms				Grass type organisms	Small bush organisms	Spineless organisms	
Designation	*(a) Enteromorpha, Ulva and Cladophora), (b) (Ectocarpus and Fucus, and (c) (Ceraminum)*	*Balanus*	*Barnacles*	*Molluscs*	*Fouling bryozoans*	*Hydroids or bryozoans*	*Hydroids or bryozoans*	*Ascidians*	*Sponges and sea anemones*
Example of typical aspect	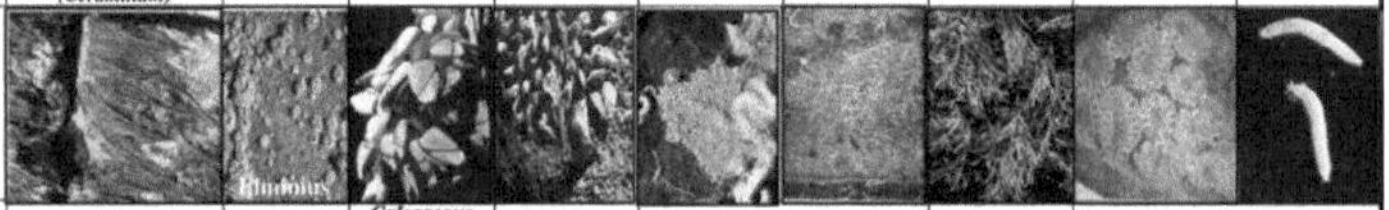								
Designation	*Green algae*	*Balanus*	*Calcareous polychaetes*	*Molluscs*	*Fouling bryozoans*		*Bryozoans*	*Ascidians*	
Example of typical aspect	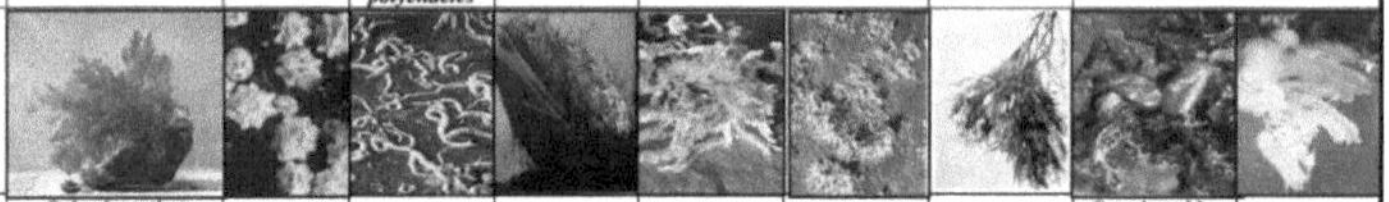								
Short description	Only plants that become attached to immersed surface: a) close to surface; b) at mid depth; and c) at depth	Attached trunco-conical or cylindrical crustaceans	Barnacles are Balanus that are fixed to surfaces via a stem	Bivalves containing a spineless animal in their interior	Calcareous incrustations that multiply from a central individual	Organisms that cover surfaces with an open grass or fur	Like bushes of several centimetres and with branches	Constituted by a spineless bag with two tubular openings or starry plates	Spineless and spongy aspect (sponges) and sea anemones

Source: Reprinted from Ref. [38]. Copyright (2007), with permission from Elsevier.

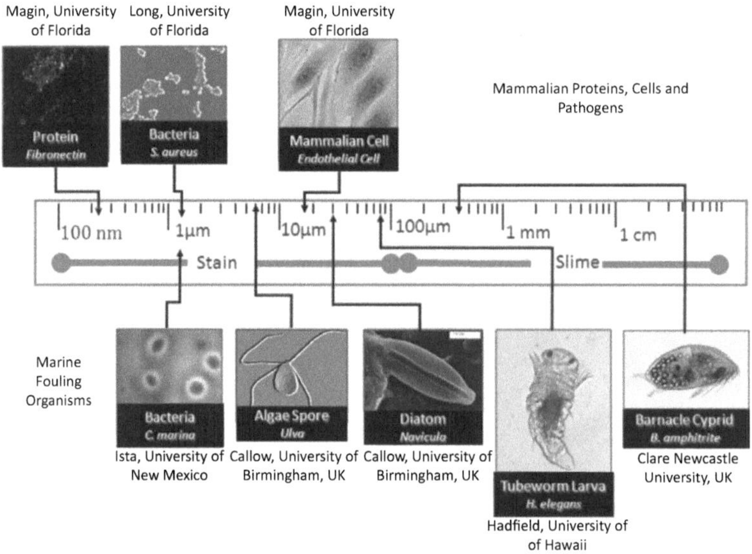

Figure 5.6 Schematic illustration demonstrating the hierarchy of fouling organisms. Cells and compounds relevant to biomedical applications are shown above the scale axis. Marine organisms are shown below the scale. Reprinted from Ref. [36]. Copyright (2010), with permission from Elsevier.

Lu et al. prepared a superhydrophobic poly(vinylidene fluoride) (PVDF) composite membrane by the surface modification of a hydrophobic PVDF membrane with the deposition of silica nanoparticles (SiO_2, ~30 nm) and 1H,1H,2H,2H-perfluorooctyl trichlorosilane (PFTS) layer [57]. The resulting SiO_2-PFTS-PVDF membrane exhibited superhydrophobicity with a surface CA of 167° ± 0.6°. The authors also examined the antifouling mechanism of the pristine PVDF membrane, a polyvinyl alcohol (PVA)-PVDF membrane, and a SiO_2-PFTS-PVDF superhydrophobic composite membrane (Fig. 5.7).

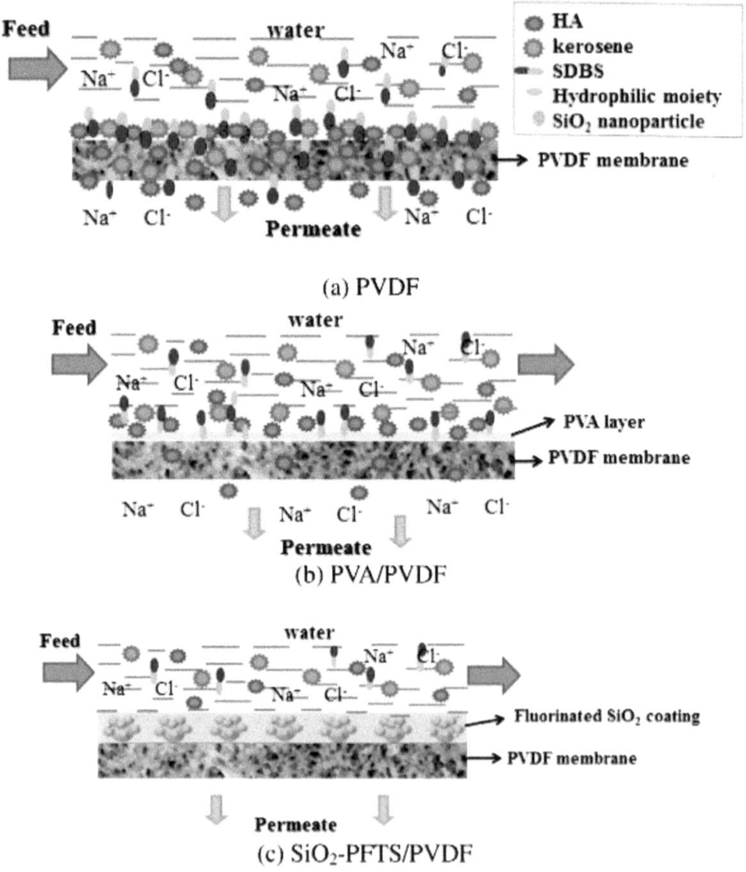

Figure 5.7 Schematic diagram of antifouling mechanism in PVDF, PVA/PVDF and SiO_2-PFTS/PVDF membranes. Reprinted from Ref. [57]. Copyright (2017), with permission from Elsevier.

The samples were placed in contact with a NaCl solution containing hydrocarbons, organic acids, and sodiumdodecyl benzene sulfonate (SDBS) surfactants. The NaCl permeation mechanism on the surfaces was examined on a hydrophobic PVDF membrane, which showed gradual changes in the surface property from hydrophobic to hydrophilic by the penetration of foulants. The superhydrophilic PVA-PVDF membrane showed strong repulsion to the surfactants in a NaCl solution and PVA could adsorb the hydrophilic moieties of SDBS in water, leading to good antifouling resistance. On the other hand, the SiO_2-PFTS-PVDF superhydrophobic composite membrane exhibited excellent resistance to the above solution, thereby preventing the permeation of a NaCl solution due to nonstick and water-repellent properties, leading to excellent antifouling performance.

The biofouling on a superhydrophobic substrate surface was also studied in seawater [58]. The authors immersed the superhydrophobic substrate in seawater for various immersion times at different sliding angles (SAs). The superhydrophobic substrates showed less biofouling in a marine environment due to the excellent water-repellent, nonstick, and self-cleaning properties (Fig. 5.8). Moreover, antifouling behavior was observed on the superhydrophobic surface for more than four weeks with a thin layer of the deposition of microorganisms compared to the pristine uncoated substrate. The SA of the substrate also played an important role in the antifouling behavior of the coated substrate. The superhydrophobic substrate maintained its surface property for more than two weeks at an SA of 0° and 90° in seawater, whereas the superhydrophobic surface was more stable (over four weeks) with an SA of 180° and also showed excellent antifouling behavior in seawater (Fig. 5.8) [58].

The superhydrophobic surface fabricated by the polymerization of dopamine in a Tris–HCl buffer solution followed by the deposition of silver nanoparticles on the polymer surface and subsequent surface modification with 1H,1H,2H,2H-perfluorodecanethiol resulted in excellent antifouling property due to the presence of silver nanoparticles on the surface of the coated matrix, which prevented the growth of bacteria and other microorganisms on their surface [59]. Furthermore, Chung et al. and Park et al. also developed silver nanoparticle–loaded superhydrophobic coated substrates, which resulted in excellent antifouling resistance of the coating materials [60, 61].

(a)

(b)

Figure 5.8 Comparison between (a) plain and (b) coated surface at 90° after 28 days of immersion. Reprinted from Ref. [58]. Copyright (2016), with permission from Elsevier.

5.2.5 Photocatalysis

Recently, superhydrophobic materials and surfaces have been used widely for photocatalytic applications because of the excellent self-cleaning property of the substrate. By a combination of superhydrophobicity with photocatalytic light responsible nanomaterials, the resulting materials can show excellent self-cleaning properties due to the easier decomposition of the deposited organic pollutant on the materials surface under the light source.

A multifunctional sol-gel coating was developed using a dense and porous self-cleaning TiO_2 films, and dense and porous multifunctional sandwiched mesoporous SiO_2 and TiO_2 films, which showed excellent photocatalytic self-cleaning behavior and antireflection properties [62]. The photocatalytic behavior of the developed substrate was assessed against the methylene blue (MB) dye solution in a custom-made UV lamp irradiation under dark conditions for 5 h. The photocatalytic degradation of MB dye was examined on the basis of a decrease in the absorption intensity at 664 nm. The developed self-cleaning and multifunctional self-cleaning surfaces did not show any photodegradation in the absence of UV-visible light exposure of the TiO_2 layer, whereas the self-cleaning coating developed using dense TiO_2 nanoparticles (18 nm thick) showed 26.7% MB degradation after 5 h.

On the other hand, the degradation of MB was increased to 47.3% and 52.0% using mesoporous TiO_2 nanoparticles (18 nm thick) with a thickness of 22 nm and 391 nm, respectively, due to the increase in the specific surface area of the coatings. These results suggest that increasing the surface area of the TiO_2 nanoparticles would show better photocatalytic activity. Further modification of the self-cleaning surface with an antireflective mesoporous SiO_2 film would enhance the degradation degree to 25%–30% more than that in the absence of antireflective mesoporous SiO_2 film. This is due to the antireflection properties of the SiO_2 film, which improves the light transmission and further enhances the photocatalytic degradation of organic dyes. The antireflection SiO_2 layer at the inner surface and the dense and porous TiO_2 layer at the outer surface with a thickness of 30 nm showed a photodegradation capacity of 51.9% and 73%, respectively. These dual advantages of self-cleaning and antireflection property of the developed film can show excellent applicability for use in various applications.

Yang et al. synthesized photocatalytic superhydrophobic hierarchical flower-like Sb_2WO_6 microspheres with a controlled morphology using a solvothermal approach [63]. The photocatalytic property of the hierarchical Sb_2WO_6 microspheres was checked against the different pH of the Rhodamine B (RhB) dye under UV-visible (500 W UV light and 300 W Xe lamp) light irradiation for

3 h. The synthesized Sb_2WO_6 microspheres exhibited excellent photocatalytic degradation behavior for RhB (553 nm), even at higher solution concentrations. Moreover, the degradation of RhB increased with increasing UV exposure time, which also depends on the controlled size and shape of the Sb_2WO_6.

Similarly, the authors also investigated the degradation behavior of RhB under visible light using Sb_2WO_6 microspheres with or without hydrogen peroxide (H_2O_2). The Sb_2WO_6 microspheres showed an excellent photocatalytic response for the degradation of RhB under visible light in the presence of H_2O_2 than in the absence of H_2O_2 with degradation percentages of ~54.7%–80.3% and ~25.7%–45.9%, respectively (Fig. 5.9). The results showed the efficient role of the electron scavenger by H_2O_2, which generates more hydroxyl groups that enhance the photocatalytic activity.

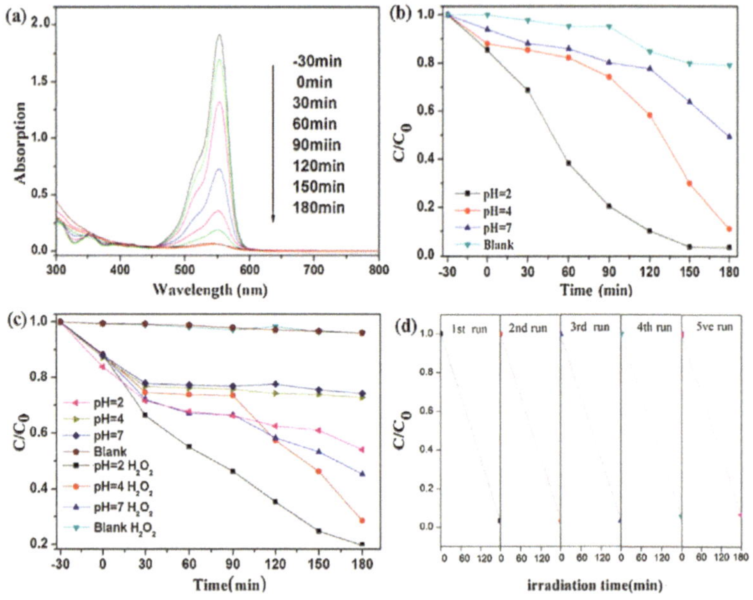

Figure 5.9 (a) Temporal evolution of the spectra during the photodegradation of RhB mediated by Sb_2WO_6 microspheres photocatalyst under under UV light irradiation. (b) Photocatalytic activities of Sb_2WO_6 prepared at different pH under UV light. (c) Photocatalytic activities of Sb_2WO_6 prepared at different pH under visible light. (d) Photostability tests over Sb_2WO_6 for the cycling photodegradation of RhB under UV light irradiation. Reprinted from Ref. [63]. Copyright (2016), with permission from Elsevier.

Furthermore, Wood et al. developed a photoconductive multifunctional poly(4-vinylpyridine) (PVP)–zinc oxide film with superhydrophobic, photoswitching, and antibacterial properties [64]. The superhydrophobic substrate was fabricated by pulse electrochemical polymerization of PVP and immersion in a zinc nitrate–dimethylaminoborane solution. The presence of zinc oxide in the fabricated substrate resulted in increased electrical conductivity by UV irradiation, which decomposed slowly upon the termination of UV exposure. On the other hand, the electrical conductivity was retained by storing the substrate under an ultrahigh vacuum (pressure < 10^{-8} mbar) for a week due to the adsorption of molecular oxygen.

Interestingly, the superhydrophobicity of the substrate can be switched to hydrophilic by UV irradiation and the superhydrophobic behavior can be retained by storing the substrate under an ultra-high vacuum. The switchability in superhydrophobicity under UV irradiation is due to the photocatalytic removal of the carbon contaminant and desorption of molecular oxygen from the zinc oxide.

References

1. Yu, L., Zhou, X., and Jiang, W. (2016). Low-cost and superhydrophobic magnetic foam as an absorbent for oil and organic solvent removal, *Ind. Eng. Chem. Res.*, **55**, pp. 9498–9506.
2. Lei, Z., Deng, Y., and Wang, C. (2016). Ambient-temperature fabrication of melamine based sponges coated with hydrophobic lignin shells by surface dip adsorbing for oil/water separation, *RSC Adv.*, **6**, pp. 106928–106934.
3. Liu, Y., Zhang, K., Yao, W., Liu, J., Han, Z., and Ren, L. (2016). Bioinspired structured superhydrophobic and superoleophilicstainless steel mesh for efficient oil-water separation, *Colloids Surf. A: Physicochem. Eng. Aspects*, **500**, pp. 54–63.
4. Si, Y., and Guo, Z. (2016). Eco-friendly functionalized superhydrophobic recycled paper with enhanced flame-retardancy, *J. Colloid Interface Sci.*, **477**, pp. 74–82.
5. Gao, Z., Zhai, X., Liu, F., Zhang, M., Zang, D., and Wang, C. (2015). Fabrication of TiO_2/EP super-hydrophobic thin film on filter paper surface, *Carbohydr. Polym.*, **128**, pp. 24–31.

6. Cao, C., Ge, M., Huang, J., Li, S., Deng, S., Zhang, S., Chen, Z., Zhang, K., Al-Deyab, S. S., and Lai, Y. (2016). Robust fluorine-free superhydrophobic PDMS–ormosil@fabrics for highly effective self-cleaning and efficient oil–water separation, *J. Mater. Chem. A*, **4**, pp. 12179–12187.
7. Zhu, J., Liu, B., Li, L., Zeng, Z., Zhao, W., Wang, G., and Guan, X. (2016). Simple and green fabrication of a superhydrophobic surface by one-step immersion for continuous oil/water separation, *J. Phys. Chem. A*, **120**, pp. 5617–5623.
8. Nagappan, S., and Ha, C. S. (2015). Emerging trends in superhydrophobic surface based magnetic materials: fabrications and their potential applications, *J. Mater. Chem. A*, **3**, pp. 3224–3251.
9. Nagappan, S., Park, J. J., Park, S. S., Lee, W. K., and Ha, C. S. (2013). Bio-inspired, multi-purpose and instant superhydrophobic lotus leaf powder hybrid micro-nanocomposites for selective oil spill capture, *J. Mater. Chem. A*, **1**, pp. 6761–6769.
10. Shang, B., Wang, Y., Peng, B., and Deng, Z. (2016). Bioinspired polydopamine particles-assisted construction of superhydrophobic surfaces for oil/water separation, *J. Colloid Interface Sci.*, **482**, pp. 240–251.
11. Li, Y., Zhang, H., Fan, M., Zhuang, J., and Chen, L. (2016). A robust salt-tolerant superoleophobic aerogel inspired by seaweed for efficient oil-water separation in marine environments, *Phys. Chem. Chem. Phys.*, **18**, pp. 25394–25400.
12. Wang, B., Liu, Y., Zhang, Y., Guo, Z., Zhang, H., Xin, J. H., and Zhang, L. (2015). Bioinspired superhydrophobic Fe_3O_4 @Polydopamine@ Ag Hybrid nanoparticles for liquid marble and oil spill, *Adv. Mater. Interfaces*, **2**, pp. 1500234 (1–10).
13. Sukamanchi, R., Mathew, D., and Santhosh Kumar, K. S. (2017). Durable superhydrophobic particles mimicking leafhopper surface: superoleophilicity and very low surface energy, *ACS Sustainable Chem. Eng.*, **5**, pp. 252–260.
14. Yu, L., Zhou, X., and Jiang, W. (2016). Low-cost and superhydrophobic magnetic foam as an absorbent for oil and organic solvent removal, *Ind. Eng. Chem. Res.*, **55**, pp. 9498–9506.
15. Lin, Y., Shen, Y., Liu, A., Zhu, Y., Liu, S., and Jiang, H. (2016). Bio-inspiredly fabricating the hierarchical 3D porous structure superhydrophobic surfaces for corrosion prevention, *Mater. Des.*, **103**, pp. 300–307.
16. Pan, S., Wang, N., Xiong, D., Deng, Y., and Shia, Y. (2016). Fabrication of superhydrophobic coating via spraying method and its applications in anti-icing and anti-corrosion, *Appl. Surf. Sci.*, **389**, pp. 547–553.

17. Xu, N., Sarkar, D. K., Chen, X. G., and Tong, W. P. (2016). Corrosion performance of superhydrophobic nickel stearate/nickel hydroxide thin films on aluminum alloy by a simple one-step electrodeposition process, *Surf. Coat. Technol.*, **302**, pp. 173–184.

18. Liu, W., Xu, Q., Han, J., Chen, X., and Min, Y. (2016). A novel combination approach for the preparation of superhydrophobic surface on copper and the consequent corrosion resistance, *Corros. Sci.*, **110**, pp. 105–113.

19. Liu, Y., Cao, H., Chen, Y., Chen, S., and Wang, D. (2016). Self-assembled super-hydrophobic multilayer films with corrosion resistance on copper substrate, *RSC Adv.*, **6**, pp. 2379–2386.

20. Wang, H., Zhu, Y., Hu, Z., Zhang, X., Wu, S., Wang, R., and Zhu, Y. (2016). A novel electrodeposition route for fabrication of the superhydrophobic surface with unique self-cleaning, mechanical abrasion and corrosion resistance properties, *Chem. Eng. J.*, **303**, pp. 37–47.

21. Li, H., Yu, S., Han, X. X, and Zhao, Y. (2016). A stable hierarchical superhydrophobic coating on pipeline steel surface with self-cleaning, anticorrosion, and anti-scaling properties, *Colloids Surf. A: Physicochem. Eng. Aspects*, **503**, pp. 43–52.

22. Wang, H., Wang, C., Gao, D., Li, M., Zhu, Y., and Zhu, Y. (2016). Heat/durability resistance of the superhydrophobic PPS-based coatings prepared by spraying non-fluorinated polymer solution, *Colloid Polym. Sci.*, **294**, pp. 1519–1527.

23. Li, J., Wu, R., Jing, Z., Yan, L., Zha, F., and Lei, Z. (2015). One-step spray-coating process for the fabrication of colorful superhydrophobic coatings with excellent corrosion resistance, *Langmuir*, **31**, pp. 10702–10707.

24. Fu, Q., Wu, X., Kumar, D., Ho, J. W. C., Kanhere, P. D., Srikanth, N., Liu, E., Wilson, P., and Chen, Z. (2014). Development of sol–gel icephobic coatings: effect of surface roughness and surface energy, *ACS Appl. Mater. Interfaces*, **6**, pp. 20685–20692.

25. Beemer, D. L., Wang, W., and Kota, A. K. (2016). Durable gels with ultra-low adhesion to ice, *J. Mater. Chem. A*, **4**, pp. 18253–18258.

26. Pan, S., Wang, N., Xiong, D., Deng, Y., and Shi, Y. (2016). Fabrication of superhydrophobic coating via spraying method and its applications in anti-icing and anti-corrosion, *Appl. Surf. Sci.*, **389**, pp. 547–553.

27. Wang, L., Gong, Q., Zhan, S., Jiang, L., and Zheng, Y. (2016). Robust anti-icing performance of a flexible superhydrophobic surface, *Adv. Mater.*, **28**, pp. 7729–7735.

28. Maitra, T., Jung, S., Giger, M. E., Kandrical, V., Ruesch, T., and Poulikakos, D. (2015). Superhydrophobicity vs. ice adhesion: the quandary of robust icephobic surface design, *Adv. Mater. Interfaces*, **2**, pp. 1500330 (1–8).
29. Wang, Y., Liu, J., Li, M., Wang, Q., and Chen, Q. (2016). The icephobicity comparison of polysiloxane modified hydrophobic and superhydrophobic surfaces under condensing environments, *Appl. Surf. Sci.*, **385**, pp. 472–480.
30. Cohen, N., Dotan, A., Dodiuk, H., and Kenig, S. (2016). Thermomechanical mechanisms of reducing ice adhesion on superhydrophobic surfaces, *Langmuir*, **32**, pp. 9664–9675.
31. Kim, P., Wong, T. S., Alvarenga, J., Kreder, M. J., Adorno-Martinez, W. E., and Aizenberg, J. (2012). Liquid-infused nanostructured surfaces with extreme anti-ice and anti-frost performance, *ACS Nano*, **6**, pp. 6569–6577.
32. Wilson, P. W., Lu, W., Xu, H., Kim, P., Kreder, M. J., Alvarengad, J., and Aizenberg, J. (2013). Inhibition of ice nucleation by slippery liquid-infused porous surfaces (SLIPS), *Phys. Chem. Chem. Phys.*, **15**, pp. 581–585.
33. Stamatopoulos, c., Hemrle, j., Wang, D., and Poulikakos, D. (2017). Exceptional anti-Icing performance of self-impregnating slippery surfaces, *ACS Appl. Mater. Interfaces*, **9**, pp. 10233–10242.
34. Yebra, D.M., Kiil, S., and Johansen, K.D. (2004). Antifouling technology: past, present and future steps towards efficient and environmentally friendly antifouling coatings, *Prog. Org. Coat.*, **50**, pp. 75–104
35. Banerjee, I., Pangule, R.C., and Kane, R.S. (2011). Antifouling coatings: recent developments in the design of surfaces that prevent fouling by proteins, bacteria, and marine organisms, *Adv. Mater.*, **23**, pp. 690–718.
36. Magin, C.M., Cooper, S.P., and Brennan, A.B. (2010). Non-toxic antifouling strategies, *Mater. Today*, **13**, pp. 36–44.
37. Chambers, L.D., Stokes, K.R., Walsh, F.C., and Wood, R.J.K. (2006). Modern approaches to marine antifouling coatings, *Surf. Coat. Technol.*, **201**, pp. 3642–3652.
38. Almeida, E., Diamantino, T.C., and Sousa, O.D. (2007). Marine paints: the particular case of antifouling paints, *Prog. Org. Coat.*, **59**, pp. 2–20.
39. Miyagawa, H., Yamauchi, K., Kim, Y.K., Ogawa, K., Yamaguchi, K., and Suzaki, Y. (2012). Fabrication of transparent antifouling thin films with

fractal structure by atmospheric pressure cold plasma deposition, *Langmuir*, **28**, pp. 17761–17765.

40. Pester, C.W., Poelma, J.E., Narupai, B., Patel, S.N., Su, G.M., Mates, T.E., Luo, Y., Ober, C.K., Hawker, C.J., and Kramer, E.J. (2016). Ambiguous anti-fouling surfaces: facile synthesis by light-mediated radical polymerization, *J. Polym. Sci. Part A: Polym. Chem.*, **54**, pp. 253–262.
41. Fujie, T., Haniudaa, H., and Takeoka, S. (2011). Convenient method for surface modification by patching a freestanding anti-biofouling nanosheets, *J. Mater. Chem.*, **21**, pp. 9112–9120.
42. Nair, A.K., Isloor, A.M., Kumar, R., and Ismail, A.F. (2013). Antifouling and performance enhancement of polysulfone ultrafiltration membranes using $CaCO_3$ nanoparticles, *Desalination*, **322**, pp. 69–75.
43. Perry, G., Coffinier, Y., Boukherrouba, R., and Thomy, V. (2013). Investigation of the anti-biofouling properties of graphene oxide aqueous solutions by electrowetting characterization, *J. Mater. Chem. A*, **1**, pp. 12355–12360.
44. Zang, D., Yi, H., Gu, Z., Chen, L., Han, Dong., Guo, X., Wang, S., Liu, M., and Jiang, L. (2017). Interfacial engineering of hierarchically porous NiTi/hydrogels nanocomposites with exceptional antibiofouling surfaces, *Adv. Mater.*, **29**, pp. 1602869 (1–7).
45. Wang, Z., and Zuilhof, H. (2016). Antifouling properties of fluoropolymer brushes toward organic polymers: the influence of composition, thickness, brush architecture, and annealing, *Langmuir*, **32**, pp. 6571–6581.
46. Grinthal, A., and Aizenberg, J. (2014). Mobile Interfaces: liquids as a perfect structural material for multifunctional, antifouling surfaces, *Chem. Mater.*, **26**, pp. 698–708.
47. Xiao, L., Li, J., Mieszkin, S., Fino, A.D., Clare, A.S., Callow, M.E., Callow, J.A., Grunze, M., Rosenhahn, A., and Levkin, P.A. (2013). Slippery liquid-infused porous surfaces showing marine antibiofouling properties, *ACS Appl. Mater. Interfaces*, **5**, pp. 10074–10080.
48. Cao, M., Guo, D., Yu, C., Li, K., Liu, M., and Jiang, L. (2016). Water-repellent properties of superhydrophobic and lubricant-infused "slippery" surfaces: a brief study on the functions and applications, *ACS Appl. Mater. Interfaces*, **8**, pp. 3615–3623.
49. Sotto, A., Boromand, A., Balta, S., Kimd, J., and Bruggen, B.V. (2011). Doping of polyethersulfone nanofiltration membranes: antifouling effect observed at ultralow concentrations of TiO_2 nanoparticles, *J. Mater. Chem.*, **21**, pp. 10311–10320.

50. Li, J., Wang, G., Meng, Q., Ding, C., Jiang, H., and Fang, Y. (2014). A biomimetic nano hybrid coating based on the lotus effect and its anti-biofouling behaviors, *Appl. Surf. Sci.*, **315**, pp. 407–414.
51. Genzer, J., and Efimenko, K. (2006). Recent developments in superhydrophobic surfaces and their relevance to marine fouling: a review, *Biofouling*, **22**, pp. 339–360.
52. Chapman, J., and Regan, F. (2012). Nanofunctionalized superhydrophobic antifouling coatings for environmental sensor applications-advancing deployment with answers from nature, *Adv. Eng. Mater.*, **14**, pp. B187–B184.
53. Junaidia, M.U.M., Ahmada, N.N.R., Leoa, C.P., and Yee, H.M. (2016). Near superhydrophobic coating synthesized from rice husk ash: anti-fouling evaluation, *Prog. Org. Coat.*, **99**, pp. 140–146.
54. Kim, Y., and Hwang, W. (2015). Wettability modified aluminum surface for a potential antifungal surface, *Mater. Lett.*, **161**, pp. 234–239.
55. Shirtcliffe, N.J., and Roach, P. (2013). Superhydrophobicity for antifouling microfluidic surfaces, *Methods Mol. Biol.*, **949**, pp. 269–81.
56. Xue, C.H., Guo, X.J., Ma, J.Z., and Jia, S.T. (2015). Fabrication of Robust and antifouling superhydrophobic surfaces via surface-initiated atom transfer radical polymerization, *ACS Appl. Mater. Interfaces*, **7**, pp. 8251–8259.
57. Lu, X., Peng, Y., Qiu, H., Liu, X., and Ge, L. (2017). Anti-fouling membranes by manipulating surface wettability and their anti-fouling mechanism. *Desalination*, **413**, pp. 127–135.
58. Cirisano, F., Benedetti, A., Liggieri, L., Ravera, F., Santini, E., and Ferrari, M. (2016). Amphiphobic coatings for antifouling in marine environment, *Colloids Surf. A: Physicochem. Eng. Aspects*, **505**, pp. 158–164.
59. Liu, Q., Huang, B., and Huang, A. (2013). Polydopamine-based superhydrophobic membranes for biofuel recovery, *J. Mater. Chem. A*, **1**, pp. 11970–11974.
60. Chung, J.S., Kim, B.G., Shim, S., Kim, S.E., Sohn, E.H., Yoona J., and Lee, J.C. (2012). Silver-perfluorodecanethiolate complexes having superhydrophobic, antifouling, antibacterial properties, *J. Colloid Interface Sci.*, **366**, pp. 64–69.
61. Park, S.Y., Chung, J.W., Chae, Y.K., and Kwak, S.Y. (2013). Amphiphilic thiol functional linker mediated sustainable anti-biofouling ultrafiltration nanocomposite comprising a silver nanoparticles and poly(vinylidene fluoride) membrane, *ACS Appl. Mater. Interfaces*, **5**, pp. 10705–10714.

62. Prado, R., Beobide, G., Marcaide, A., Goikoetxea, J., and Aranzabe, A. (2010). Development of multifunctional sol–gel coatings: anti-reflection coatings with enhanced self-cleaning capacity, *Solar Energy Mater. Solar Cells*, **94**, pp. 1081–1088.

63. Yang, C., Yang, X., Li, F., Li, T., and Cao, W. (2016). Controlled synthesis of hierarchical flower-like Sb_2WO_6 microspheres: photocatalytic and superhydrophobic property, *J. Ind. Eng. Chem.*, **39**, pp. 93–100.

64. Wood, T. J., Hurst, G. A., Schofield, W. C. E., Thompson, R. L., Oswald, G., Evans, J. S. O., Sharples, G. J., Pearson, C., Petty, M. C., and Badyal, J. P. S. (2012). Electroless deposition of multi-functional zinc oxide surfaces displaying photoconductive, superhydrophobic, photowetting, and antibacterial properties, *J. Mater. Chem.*, **22**, pp. 3859–3867.

Summary and Outlook

Organic-inorganic nanohybrids have excellent properties due to a combination of the properties of organic and inorganic materials in a single material. These technical advantages of organic-inorganic nanohybrids are useful for applying the materials in diverse fields. Moreover the growth of nanotechnology over the past few decades has also increased the advantages and uses of organic-inorganic nanohybrids with huge financial support for wider nanotechnological approaches. In particular, organic-inorganic nanohybrid materials are used in various energy, environmental, and biomedical applications.

Currently, organic-inorganic nanohybrids are used to develop many commercial products. Moreover, organic-inorganic nanohybrids are used in the synthesis and fabrication of materials with varying surface properties, such as hydrophilic, hydrophobic, superhydrophobic, ominiphobic, amphiphobic, and superamphiphobic coatings. On the basis of their surface properties and nature, these mateirals are used in a wide variety of applications, such as oil sorption and separation; photocatalysis; corrosion and scratch resistance; anti-icing, antifouling, antireflection coatings; sensors; biomedical applications and drug delivery; organic dyes; and metal ion adsorption.

This book especially covers the synthesis, fabrication, and applications of hydrophobic and superhydrophobic organic-inorganic nanohybrids. Because of their partial and complete water-repellent properties, hydrophobic and superhydrophobic organic-inorganic nanohybrids used for the development of hydrophobic and superhydrophobic coatings can be applied easily to various substrates to improve the surface properties of the coated substrates.

This book will be useful for readers in interdisciplinary areas of research.

The present and future needs of the synthesis, fabrication, and applications of organic-inorganic nanohybrids are to develop environmentally friendly and solvent-free materials. Currently, some innovative studies have been carried for the development of solvent-free and environmentally friendly organic-inorganic nanohybrids with hydrophobic and superhydrophobic properties.

Environmentally friendly, solvent-free organic-inorganic nanohybrid coating materials are expected to be a major focus in the future to reduce the production costs and consumption of organic solvents and be used on any substrate.

Index